Alfred C. (Alfred Cheatham) Stokes, Alfred Cheatham Stokes

Microscopy for Beginners

Alfred C. (Alfred Cheatham) Stokes, Alfred Cheatham Stokes

Microscopy for Beginners

ISBN/EAN: 9783744686310

Printed in Europe, USA, Canada, Australia, Japan

Cover: Foto ©berggeist007 / pixelio.de

More available books at **www.hansebooks.com**

MICROSCOPY FOR BEGINNERS

OR

COMMON OBJECTS
FROM THE PONDS AND DITCHES

By ALFRED C. STOKES, M.D.

ILLUSTRATED

"The microscope is not the mere extension of a faculty; it is a new sense"

"The microscope, frequently and intelligently used, makes nature pellucid"

NEW YORK
HARPER & BROTHERS, FRANKLIN SQUARE
1887

CONTENTS.

CHAPTER VIII.

CHAPTER IX.

CHAPTER X.

CHAPTER XI.

CHAPTER XII.

ILLUSTRATIONS.

INTRODUCTION.

To the beginner in the use of the microscope, indeed to the beginner in the study of any department of natural history, the name of the specimens found is of the first importance. It is the key that opens the door to further knowledge, and until it is obtained the beginner is helpless; the books are closed to him, all conference with others in reference to the object or specimen is impossible, and, in many, a budding interest that might otherwise bloom and bear fruit is crushed and destroyed. The first question asked is always, "What is it?" and unless the questioner has a kind and experienced friend to whom he can take the specimen, or a book of common objects from which the names of ordinary natural history materials can be ascertained, the question is too often unanswered, and the beginner soon loses his relish for the unknown in nature, because to him it always remains the unknowable.

In England innumerable little hand-books in all departments of natural science are within the reach of every reader, even the least wealthy. They are written in an attractive style, they are usually accurate as far as they go, and they aim to describe the common objects to be found in the green lanes and the woods, the waters of the ponds and streams, and the shallow bays and inlets of the sea, so that any one with the least inclination towards the study of the teeming world of animal and vegetable life can, at slight expenditure of time, labor, and money, learn the names of the commonest things

A*

surrounding him. Such books, if correct and helpful, are worthy of all praise. That there is a desire for them, even in this fair land of ours, is evident by their importation, and their appearance on the counters of the booksellers and the shelves of the public libraries. But they are seldom adapted to our needs. Their descriptions are commonly too general and diffuse, their writers pay more attention to literary style than to the imparting of definite information, and the text too often bears internal evidence of having been made to suit certain pictures owned and necessary to be utilized by the publisher. That similar and better books on the life in American fields and streams, and American sea-shores, are so few is much to be regretted. There ought to be small and untechnical hand-books adapted to "all capacities, even the meanest," as our forefathers used to put it, and in all departments of animal and vegetable life ; books in which the beginner could learn the names of things. "I do beseech you, what is your name?" is the oft-asked question, not only by the beginner in the use of the microscope, but by the more advanced student in other departments as well. Emerton's "Life on the Sea-shore," and his "Structure and Habits of Spiders;" Hervey's charming "Sea-mosses," Gray's "How Plants Grow," Romyn Hitchcock's "Synopsis of the Fresh-water Rhizopods," Jordan's "Manual of the Vertebrates," are delightful books that approach the ideal nearer than any others published in this country ; indeed, there are no others. There is so much for our learned scientists to do in this comparatively unexplored land of ours, that they may have no time to stoop and lend a hand to those who would like to enter a little way into the attractive world of science, from which faint but pleasant rumors occasionally come. They are all courteous and communicative when personally approached, but what boy or other young person with an inclination towards "bugs and things" would be willing, or, indeed, would know how to seek aid from these celebrated men? And if the student is alone in a country place where Nature smiles her sweetest, but where there are no libraries and no human being to consult, except, perhaps, "the minister," how then shall he learn the name of the flower, the stone, or the bird that attracts his attention? "The minister" is usually poor authority on such subjects, and the boy, after wondering and investigating in an awkward and boyish fashion, soon gives it up, when he might have become a lover of nature, and perhaps a lover of something even better than nature. The "Agassiz Association," with its clubs and chapters and auxiliary natural history societies, is doing much good in awakening a desire in its young members to know something of natural science, and in doing something to help the young investigators. Yet it can do but little. The workers must depend upon themselves, and the books, of which there are so few adapted to their needs.

The microscope is every day becoming a more familiar instrument to the young. There is a growing interest among the boys and girls, even among those of a larger growth, in the little things of the world, and the number of so-called microscopists is rapidly increasing. But the possessor of an instrument looks at the two or three mounted objects supplied by the dealer, and then wonders if this is all, and if this is the only foundation for the charming stories he has heard of the charming things to be seen with the microscope.

"Will you tell me where I can find a book that will help me to know a microscopic plant from a microscopic animal, and teach me how I can best collect them?" is a question that has often, in some shape, been asked the writer, and has as often remained unanswered, for there is no book on common American microscopic objects. It is only possible to direct the questioner to the ditches and the ponds, and to wish him a success that is almost hopeless. In any event, the beginner naturally, and almost instinctively, goes first to the water for his microscopic objects, probably because he has heard so much about the "animalcules" there. His first examination bewilders him. There is so much life and motion and color, there are so many strange forms; but where shall he turn for help?

Since our illustrious scientists have not offered to help him, the writer, who is only a beginner himself, and who makes not the slightest pretensions, has sympathized with the inquirers whom he has been compelled to turn away unsatisfied when they have come for printed help in their microscopical work, and this little book is the result. It claims no literary merit; it makes no scientific pretensions. Its only aim is to help the beginner to ascertain the names of some of the common microscopic creatures, both animal and vegetable, with which the fresh waters of the land are filled, and it tries to do so in the simplest and most direct way, leaping scientific hedges and trampling on scientific classification in a manner that will dismay the learned botanist and zoologist. But the botanist and zoologist have weighty books that delight their souls, so why should not the beginner with a microscope have a book to help him to the names of the commonest aquatic objects, and, it is hoped, delight him by smoothing the path that leads to them? The writer will not be greatly troubled if the learned botanist and zoologist do not like this little book, provided the beginner in the use of the microscope approves it and finds it helpful.

It relates almost exclusively to aquatic objects. One reason for this has already been mentioned. Another and more potent one is, that even the beginner knows, in a general way, what he is looking at when he magnifies the common objects of the land, but the microscopic creatures from the water are so truly microscopic, the observer must so often go fishing on faith, and only know the contents of his net by faith and imagination until he can examine his collection drop by drop with the microscope, that he is lost at the start unless he has a book to help him, which this one hopes to do. But is it necessary to say that the following pages do not contain notices of everything to be found in the ponds and ditches? The beginner will capture many objects which he will not find described here. It is not possible that the matter should be otherwise. The waters are crowded with life, and it is only the commonest objects and those most frequently found, that a little book of this kind can attempt to include.

The descriptions of those few have been made as plain as possible. The writer has seldom allowed himself to "fall into poetry," although often sorely tempted. The keys or analytical tables so freely scattered through the pages have been purposely made as artificial as they could be. They use the most conspicuous external characters without regard to scientific classification, and without regard to any result but one only—to help the beginner find the name, at least the generic name, of his specimen. If this is accomplished the book will have attained its purpose. The method of using the keys is explained on page 70.

Finally, to the beginners in the use of the microscope, for whom the book has been prepared, the writer would say, as has so often been already said: There is no royal road. The mother-bird finds and brings the food, but even the youngest nestling opens its own mouth.

MICROSCOPY FOR BEGINNERS.

CHAPTER I.

THE MICROSCOPE AND ITS PARTS.

Simple and Compound Microscopes.—Pocket-lens. —"Craig Microscope."—"Excelsior Microscope."—Watch-maker's Glass.—Coddington Lens.—How to Focus a Simple Lens.—Parts of the Compound Microscope. — Draw - tube. — Eye - pieces. — Society - screw.—French Triplets.—Objectives.—Selecting Objectives for the Beginner.—Coarse Adjustment.—Focussing.—Fine Adjustment.—The Stage.—Diaphragm.—Mirror and Bull's-eye Condensing-lens.—Preparing the Object.—Thin Cover-glass.—Cells. —Cement.—Dry Mounting.—Needles.—Dipping-tube.—Bunsen Burner. —Evaporation from beneath the Cover. —Life-slide.— Growing-cell. —Air-bubbles. —Drawing. —Camera Lucida and Glass Reflector.—Micrometer.—Measuring the Object.—To ascertain the Magnifying Power.—Collecting-bottles.—Books and Magazines for Reference.

MICROSCOPES are compound or simple: compound when they consist of two or more glasses, one or more being near the object to be examined, and one or more near the eye of the observer; simple when they consist of but one double-convex lens to be held near the object, or of two or more lenses that can be used singly or all at the same time. When thus used in combination, the two or three simple lenses are not only placed close

1

to each other, but close to the object, the combination
acting as if it were a single lens, the magnifying power
being much greater than that of but one glass, and the
distance from the object much shorter when in focus.
In the compound microscope the lenses near the eye
magnify the image formed by the lower glasses, and
that image is inverted, the upper side of the object then
appearing to be the lower, the right-hand side the left,
and the left-hand the right. In the simple microscope,
however, the image is not inverted; and in those forms
where two or three lenses are combined, the effect is the
same as though one glass of great magnifying power
were used. But separate the lenses so that the upper
shall magnify the image produced by the lower, and you
have a simple form of compound microscope. In the
simple microscope we see the object itself, in the compound we see the enlarged image of the object.

As a simple microscope does not seem to invert and
reverse the object, and because the distance between the
two is long when a low-power glass is in focus—that is,
when the glass is in such a position that the magnified
object looks clear and distinct to the eye—it is always
used for the examination of a flower, the surface of a
piece of bark, a stone, an insect, or any other specimen
of considerable size, or one that is visible to the naked
eye, more extended study being reserved for the compound instrument at home. A simple microscope, a
"pocket-lens" as it is often and preferably called (Fig. 1),
is really indispensable to every one who has a taste for

nature studies, and a desire to know somewhat of the beauties hidden from our unaided vision; for the simplest glass shows the student unimagined charms in the petal of a flower, the sand he walks on, and in the green scum that floats on every summer pool and disgusts him until his

Fig. 1.—A Pocket-lens.

little lens reveals its purity and grace. It is always ready for the examination of anything picked up in the fields or woods, it is small, and it is easily carried in the pocket. It can be obtained in a great variety of shapes, so far as the frame that holds the lens is concerned; it can be had with but one glass, or with two or three of various powers, to be used alone or combined; it can be bought with a large lens of low power in one end of the frame, and a smaller glass of higher power in the other. But whatever form the beginner selects, he must remember that the larger the simple lens the lower, as a rule, will be the magnifying power, and the longer the working distance, or the space between the glass and the object when in focus; and the smaller the lens the more convex it will be, the greater power it will have, the shorter the working distance, and the less of the object it will show at one view, and consequently the more troublesome it will be to use. The beginner is advised to purchase a good pocket-lens with a working distance, or "focal length" as it is sometimes rather incorrectly termed, of one or one and one-half inches. This is all that is really needed for the examination of botanical specimens and the thousand and one objects that attract the attention on every summer ramble.

The writer personally dislikes the combination pocket-lens formed of two or three separable glasses. If but one lens of the combination is wanted for immediate use, the entire number must be pushed out of the thick and awkward case, one must be selected and separated, for the perverse thing usually comes out of the pocket upside down, and it is of course desirable that the highest-power glass shall be next to and nearest the object, while those not needed are turned to one side, making a series of operations that take time, both hands, and considerable patience if you are anxious to examine the specimen. Your companion will have finished the work with the single glass, and will be telling you how the object looks, before your complicated affair is ready to begin, provided you are not wise enough to have avoided the combination pocket-lens. And if the whole number is used at once, the working distance is usually so short that the observer's head or hat-brim shuts off most of the light, so that the object can be seen with difficulty, and a very little of it at that. To see at one view so small a portion as these high-power combinations always show, and to be compelled to pass the lens over so many little parts before an idea of the whole surface can be obtained, is, to say the least, not satisfactory; unless the observer is familiar with the entire object, and the relation and arrangement of all the parts, a low-power pocket-lens is the most useful, and the one to be recommended.

The reader perceives that this matter of short focus is an important one; indeed the usefulness of the pocket-lens to a great extent depends upon it. Reject without hesitation the simple lens whose focus is so short that it must be held almost in contact with the object.

Not long ago a rather expensive instrument called the "Craig Microscope" was extensively advertised, and sold as a remarkable thing. The lens was a small globule of glass fastened to a glass plate, to give it a flat under-surface, and mounted in a brass ring, the whole being supported on an upright brass tube with a plane mirror at the lower end. It was not a compound microscope, but a simple lens with mirror attachment. The object to be examined was suspended from the flat surface of the glass in a drop of water, the focus being so short that it was at the front of the lens, so that nothing could be looked at unless it was adherent to the glass. No mounted object could be satisfactorily studied; to examine the parts of a flower was impossible, and even when a drop of water was suspended from the lens its contents were distorted almost beyond recognition.

The "Excelsior Microscope" makes no false pretences. The instrument consists of a small box which is a receptacle for all the parts when not in use, and a support when the steel rod is elevated to receive the combination pocket-lens and the stage, on which the object is to be placed, a small mirror in the front of the box reflecting the light to the object from below. A great fault is the absence of weight in the instrument. At the least touch it moves, the light reflected from the mirror is lost, and the object is consequently left in semi-obscurity. It is intended chiefly for the dissection of flowers, grasses, or large insects, and fairly answers the purpose if the observer desires to have both hands free, and cares to screw the box to the table. But it is no better than a good pocket-lens, which, with very little trouble, can be attached to an upright rod and be used for dissections; in some respects it is much less valuable. The three lenses supplied can be used as a single one or combined. The former is good, the combination of two is not seriously objectionable, but the focus of the united three is, to the writer's eye, only five-sixteenths of an inch, a distance, aside from the small field of view, that effectually prevents its use as a dissecting microscope. With the lowest-power lens six letters of the type used in this book can be seen, the focal distance being one and one-fourth inches; with two lenses combined, four letters, with a focal length of about one-quarter of an inch; and with the three glasses only one letter is visible, the focal distance being five-sixteenths of an inch, when tested by the writer.

A "watch-maker's glass," which is sometimes seen on the microscopist's table, is a simple lens mounted in a short horn or rubber tube, so arranged that it can be held to the eye by the contraction of the muscles of the cheek and brow, while both hands are used for the manipulation of the object. It can be obtained of various powers and focal lengths, but it is scarcely

desirable. The prolonged contraction of the facial muscles necessary to keep it in place is very fatiguing, and
the vapor always evaporating from the front of the
eye being confined within the tube is sure to condense
on the lens and obscure the object. Everything a watch-
maker's glass will do, a good pocket-lens will accomplish.

A "Coddington lens" is admirable in many respects.
Its magnifying power is great, the image it forms is excellent, the field of view is good, but the focus is usually
unpleasantly short. This, aside from its cost, is its only
objectionable feature. It is named after the gentleman
who first brought it to the notice of the opticians, and
not, as it should have been, after Sir David Brewster,
its inventor. It consists of a sphere of glass with a deep
groove cut around its centre, and filled with a black cement, which acts as a diaphragm to cut off certain rays
of light whose presence and action would be undesirable, as they would interfere with the formation of a clear
and sharply outlined image.

The reader may be surprised to learn that there are
people who do not know how to focus a lens. I have
seen such persons take the glass as if they were afraid
of it. They extend it towards the object in a hesitating
way, move it about irregularly for a few moments, throw
back the head, look cross-eyed, and say, "Oh yes; I see.
How beautiful! And how very queer it looks!" I
once offered a lady an opera-glass, which she put to her
eyes and never touched the adjustment wheel that alters
the length of the tubes and focuses the lenses on the

actors; and when she returned it she said, "Thank you. I don't like it much; I can see a good deal better without it."

To "get the focus" it is not really necessary to close one eye, although that is usually done. If both eyes are open, the one looking through the lens becomes so interested that the other sees nothing; or, if you prefer, you may say that the brain becomes so interested in contemplating the image formed on the retina of the eye examining the magnified object, that it fails to note the retinal impressions of the other. But if one eye must be closed, it can be done, after very little practice, without clapping your hand over it. This applies equally well to the use of the compound microscope.

To focus a pocket-lens hold the object to be examined in the left hand, and while looking through the glass raise and lower it with the right hand until the magnified object appears clear and distinct, the outlines sharp, and without a fringe of color, and the surface rough or smooth, rounded or concave, as it indistinctly appears to the unaided eye. The focus cannot be obtained without this experimenting every time the glass is used. A good plan is to place the lens nearer the object than you know to be necessary, but always without allowing the two to come in contact, and then to slowly raise the glass until the image is distinct, when it will be focussed. Keep it steadily in that position and study the object.

The compound microscope (Fig. 2) consists of the stand, the eye-piece, and the objective, although the word, as commonly used, refers to the entire combination of brass, with or without the magnifying-glasses.   But

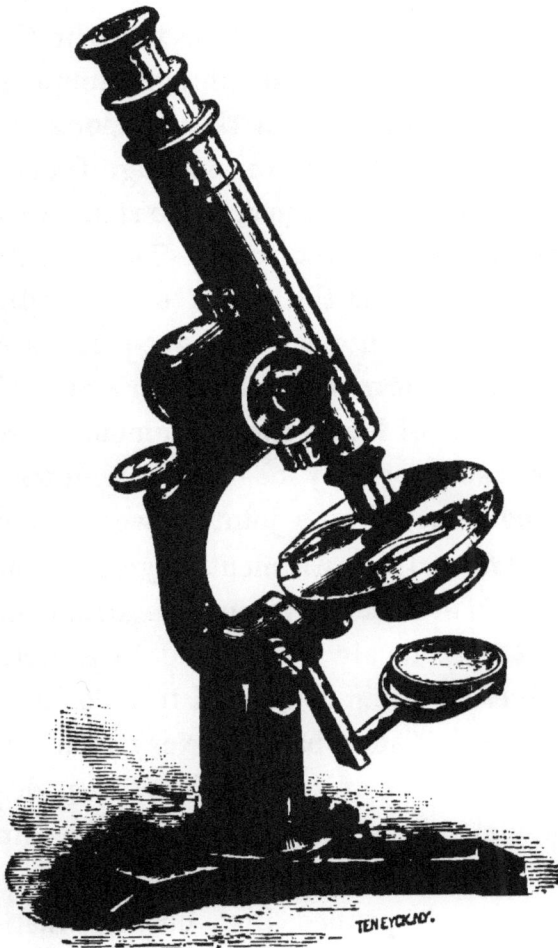

Fig. 2.—A Compound Microscope.

without the objective the microscope is only the "stand," and is practically useless.   The stand alone generally includes the tube or microscope body, the eye-piece,

1*

formed of two lenses at the opposite ends of a short tube inserted into the upper end of the body, the arm supporting the body, the stage on which the object is placed to be examined, the mirror to light the object, a movable circular plate, or diaphragm, immediately beneath the stage, and the foot that supports the whole. The addition of the objective, or magnifying-glass, at the lower end of the body, makes the stand a compound microscope of the simplest form. The objective is so named because it is near the object to be examined when the microscope is in use, and the eye-piece is so called because it is then near to the observer's eye. Without both of these sets of lenses the instrument is useless.

The arm and foot may be made of either brass or iron, and there should be a joint between them so that the upper parts of the instrument may be inclined. The cheapest stands are made without this arrangement, and they must therefore always be used in a vertical position, the observer being compelled to hold his head and body in a way that soon becomes very wearisome. An iron arm and foot are quite as useful as if made of brass, but no stand should be selected without the joint for inclination. Brass looks better, and is much more expensive than neatly japanned iron, but is practically no more useful.

The body should be about ten inches long. In the less expensive stands it is often made in two parts, the upper tube sliding within the other, so that when it is drawn out to its full extent the entire body will then be the proper length to obtain the best results from the objectives. In such a stand, when the inner, or "draw," tube is pushed down, the microscope will have the lowest magnifying power obtainable with the eye-piece and objective then in use; when fully extended, the power of the objective will be greatly increased, so that by varying the length of the body by the use of the "draw-tube," many different magnifying powers may be obtained from one objective. In some cases this arrangement may be useful; it is at least not entirely objectionable, neither is it very convenient. Stands with an undivided body ten inches long—the standard length—also often have a draw-tube by means of which the body can be enormously lengthened and the magnifying power enormously increased, but usually with a loss of some good qualities in the image. The addition is occasionally useful, but it is not necessary. If the reader selects an instrument with a body of the standard length, and he finds that it is without a draw-tube, he need not be troubled. The stand will be as valuable without as with this secondary part.

The eye-piece consists of two lenses at the opposite ends of a short brass tube divided internally by a diaphragm. The lens nearest the observer's eye when the instrument is in use is the "eye-glass," the one at the opposite extremity the "field-glass." The price of the stand usually includes one or more eye-pieces. If but one is supplied, it will generally be the lowest power, the two-inch or "A;" if two, the one-and-one-half or

2

one-inch, often also called "B" or "C," will be added.
Opticians also make $\frac{1}{2}$, $\frac{3}{4}$, $\frac{1}{8}$, and even $\frac{1}{16}$ inch eye-pieces,
most of which are for special kinds of microscopical
work, their magnifying power being enormous and the
result almost worthless; indeed, these very high power
eye-pieces are usually to be avoided. On no account
should they be selected by the beginner in microscopy.
Every purchaser of a stand should insist upon having
the two-inch, if he can have but one, as it is always useful, and is all he will need for a long time, or until he
desires to use an eye-piece micrometer for the meas-
urement of microscopic objects, when he can add the
one-inch, or "B," ocular to his stand.

The lower opening in the body always carries a screw
to receive the one on the upper end of the objective.
Several years ago the size of these screws varied widely
in stands and objectives of different makers, so that if
the student desired an objective of different make from
those accompanying his instrument, he was forced to
buy a little piece of apparatus called an adapter, one end
of which was made to screw into the microscope body,
the other to receive the objective. At the suggestion,
however, of the Royal Microscopical Society of London,
all objectives and stands now have screws of the size
recommended by that society, and therefore called the
"society-screw." Only the very cheapest stands of the
present day, or those having the least value as instruments for serious investigation, are without this screw,
and they are usually supplied with what are termed

French triplets. These are miserable lenses that should always be shunned, as they will do the observer more injury than much time can remedy.

It is true that before the optician, especially before the American optician, began to make really good objectives at moderate cost, these French triplets were extensively used, and are said to have done some good work. But at what expense? Not at the expense of any great amount of knowledge or skill in their manufacture, for the lenses were ground, mounted singly, and then combined in an experimental way: two or three were selected at random from a basketful, screwed together, and examined on a microscope. If the result was considered satisfactory, all was well; if not, one or more of the lenses was replaced by others also selected at random, and the experiments were continued until the objective was considered passable and salable. Such, at least, is the credible story. Their expense, therefore, was not in the making; it was in the imperfect image, in the great loss of light, in the injury to the eye due to the strain caused by the absence of sharpness and brilliancy characteristic of the image formed by even low-priced American objectives, and in the time wasted while unconsciously forming erroneous conclusions from the objects so imperfectly seen. The writer is somewhat emphatic on this point. He knows whereof he speaks, for he began the use of the microscope with French triplets, and employed them for years, because he was ignorant and had no teacher. What the cost was to him he knows only too well. To the young student who longs for a microscope, I am almost tempted to say, if you cannot afford the cheapest suitable low-power American objective, if you must have the ordinary French triplet or none, take none. It is a hard fate, but is not life itself hard? Fortunately, however, these inferior commercial lenses are not extensively in the market at the present day. Yet the purchaser of a microscope already fitted out with objectives, should inquire whether he is buying French triplets. If so, then as his experience, knowledge, and skill increase, so will his dissatisfaction increase. An intelligent boy had been using these poor lenses for some time, and doing work that, under the circumstances, was commendable, when he for the first time looked through a good low-power (one-inch) objective. After a momentary examination, he glanced at me in a wondering way as he said: "How beautifully bright and clear it looks! My microscope is different. I think it needs cleaning!"

Modern objectives are the result of the most consummate skill of the accomplished optician. There is no chance work in his methods. Every curve is mathematically exact, and is calculated and positively known before the glass comes to the grinding-tool. Objectives are usually a combination of several lenses, but the union is not accidentally perfected. The maker's knowledge of abstruse optics tells him the precise result to expect from the combination of lenses of certain forms made from glass of a certain chemical composition. He

is the master; his objective is a masterpiece. The owner of a good objective must not treat it carelessly. He
should treasure it, for it is not a common thing. When
not on the stand in use, it should be kept in the brass
box supplied for that purpose, and it should never be
left on the stand when not in actual employment.

That part of the brass mounting of the objective
which bears the screw is the back; the opposite end that
shows a small flat surface of glass is the front, or, as it is
often styled, the front lens. The glass of this part is
soft and easily scratched, therefore take care not to
let it touch anything hard; especially avoid any gritty
substance, or dirty rag that may hold a minute particle of sand or hard dust; and never touch it with the
fingers, as the oily exudation from the skin will soil
it and interfere with the clearness and beauty of the
image. If the front lens becomes accidentally stained,
or soiled by long use, the objective should be sent to its
maker, who can clean it without the great risk that its
owner would expose it to if an attempt should be made
to wipe the glass. If fine dust adheres too closely to
be dislodged by the breath, ravel out the edge of a piece
of very clean old linen or muslin, and with the fringe
thus obtained gently sweep the surface.

When the objective is to be taken from its box, un-
screw the cover and tip the lens into the palm of the
left hand, supporting it with the fingers; pick it up
with the thumb and finger of the right hand against the
sides of the tube or brass mounting, and it will be ready,

when reversed, to be screwed to the stand. If it is not to be returned to the box immediately after use, as will often happen if the student has more than one, and he desires to examine the object with another power, stand it on its screw end on the table, and to protect it from dust invert its box over it. The latter can be lifted off in a moment, and the objective will then be ready to be picked up as before.

What objectives should the beginner select? If possible, he should have two, a low and a moderately high magnifying power. If unable to purchase both at once, let him by all means first take what is called the one-inch objective; if he can also buy a high-power, the $\frac{1}{4}$ or $\frac{1}{5}$ will be the proper glass. But for this he can wait. There is so much to be examined with the one-inch objective that, for a long time, he will scarcely feel the need of another. The inch, if properly selected, need not be expensive, but it should be a good and satisfactory glass, not only at the outset, but when the student becomes an expert microscopist; it will then always be useful. Such objectives are made by several American opticians, and included in what they call their "Students' Series." When in focus, the distance between the front lens and the surface of the object—the "working distance"—is large, so there will be no trouble in using it; and with the two-inch, the "A" eye-piece, the magnifying power will be about forty-five diameters, or a little more than two thousand times.

After the student has been using the one-inch objective for some time, and his eye has become educated, he will begin to catch glimpses of minute objects beyond the ability of the low-power glass to properly exhibit. Then he will wish for something more, so that he can look deeper into the little things of nature. What shall it be? The opticians make $\frac{1}{8}$, $\frac{1}{10}$, and even $\frac{1}{50}$ inch objectives, which magnify enormously, cost frightfully, and can be successfully used only by accomplished microscopists on large and first-class stands. To the beginner, even after considerable experience with the low-power, any objective higher than the $\frac{1}{4}$ or $\frac{1}{5}$ will be useless. With these glasses he will be well equipped for quite extensive microscopical study, until he is ready to undertake original work in some unexplored department of science, or in some partially neglected corner, of which there are many in every scientific field, however well cultivated. Like the one-inch, the $\frac{1}{4}$ or $\frac{1}{5}$ will always be useful. As the observer's eye becomes better educated, when it learns, as it will, to see minute parts of delicate objects, which at the start were entirely overlooked, the high-power objective will not be thrown aside, the student will not become disgusted with it as he would with a high-power French triplet, but his quickened sight will again catch glimpses of beauty to be examined, and mystery to be unravelled, which are beyond the power of his best objective, and he will almost unconsciously have advanced another step.

Personally the writer prefers the $\frac{1}{5}$ inch objective to the $\frac{1}{4}$, and such a glass need not be expensive to be good (several opticians' "Students' Series" include them), the working distance is not too short, or need not be, and with the two-inch eye-piece it will give a magnifying power of about two hundred and fifty diameters.

"The coarse adjustment" is the expression usually applied to the rapid movement of the body produced by turning the large milled heads, one of which is on each side of the instrument. It is used in focussing, that is, in obtaining a distinct image of the object when seen through the eye-piece and objective. The image then appears surrounded by a circle of light called the "field of view," or simply "the field." Very few, except the small, vertical "boys' microscopes," and some of the cheapest and least desirable American or English stands, are without the coarse adjustment. Occasionally a stand will be seen in which this part is replaced by a broad, cloth-lined, or tightly-fitting collar, through which the body slides, the movement being made by hand. This is very unsatisfactory, and such stands should be avoided, if possible, as, sooner or later, the body is sure to be suddenly pushed too far down, the objective then coming in contact with the object: an accident to be always guarded against with the greatest care, as the objective, or the object, or both, may be injured. If the object is destroyed it may possibly be replaced, but a scratched or broken objective can be remedied only by buying a new one. Of course the microscope body may, by a careless student, be forced against the object by the use of the milled heads, and equally, of course, a

man may fill his stomach with gravel-stones or powder-
ed glass; but no sane man will so maltreat that organ,
and no sane microscopist will so maltreat his objective
as to drive it against the object on the stage when the
risk is so great.

The only proper way to use the coarse adjustment is
to *always focus upward.* When the object to be examined has been placed on the stage, and the light from
the mirror is properly arranged, the microscope body,
with the eye-piece and objective, is racked downward by
means of the milled heads until the front of the objective almost touches the object, the observer carefully
watching that they do not come in contact. Then place
the eye at the eye-piece, and nothing will be visible except the brightly illuminated field of view; but, while
looking into the microscope, slowly raise the body until
the image appears sharp and clear, in other words until
the objective is focussed. It makes no difference whether the distance between the front lens, when focussed,
and the object is two inches or the one-hundredth part
of one inch, always rack the objective down while you
are looking *at* it, and focus upward while you are looking *through* it. This is the single rule that must never
be forgotten. It has been said in a joking way, "that
nothing will throw a microscopist into a chill more
quickly than to see a friend look into his microscope
and *focus down* with the coarse adjustment." Yet men
who ought to know better have been seen to do this
reprehensible thing.

2*

In the older stands a single small milled head will be found on the front of the body near the lower end, just above the society-screw. In more recent stands it will be on the arm at the back of the instrument. This is the "fine adjustment screw;" and although it adds somewhat to the cost, it should always be on the stand if the purchaser desires to use even moderately high-power objectives. For low-powers it is not necessary. The fine adjustment screw is so made that by turning its milled head the objective, if the adjustment is at the front, or the entire body, if it is at the back, is slowly raised or lowered. When the high-power objective has been imperfectly focussed by racking the body *upward*, it seldom happens that the image is as distinct as is desirable; therefore the microscopist, by a few gentle turns of the fine adjustment screw, raises or lowers the objective, until the magnified image has its outlines as sharply defined as the figures in the best steel engravings. With the one-inch objective, or others still lower (two, three, or even four inch), the focus can be accurately obtained by the coarse adjustment alone, but with the $\frac{1}{4}$ or $\frac{1}{8}$ the fine adjustment must always be used.

It is a great mistake made by some who ought to know better, to try to examine an object not distinctly in focus. In such cases the strain on the eye is severe and injurious, while the pleasure of examining the preparation is much lessened. The changes made for the better by a few delicate touches of the fine adjustment can be appreciated only when seen. Always try to have the image as distinct as possible. If in doubt as to the focus, after obtaining what seems to be a moderately good appearance, give the fine adjustment a turn or two one way or the other, noticing whether the image becomes sharper in outline and clearer in its general aspect, or whether it grows cloudy and indistinct. If the last, the focus has not been improved, and was probably correct at first. A very little experience will make the beginner an expert in this important matter.

The stage, on all but the largest and most expensive instruments, is a square or circular piece of thin metal, with a large central circular opening for the passage of the light from the mirror. Sometimes the metal stage has a glass plate made to slide over it easily. This is a convenience and a desirable luxury, but it is by no means necessary. The strip of glass that bears the object to be examined can just as readily be slipped about under the objective by the fingers directly, as it can be if supported on this movable glass stage. These finger movements require a little practice, but the student will so soon become accustomed to them that he will change the position of the object without consciously thinking of it, and his touch will become so delicate that he will be able, with the slightest pressure, to move the object for a distance so small that it would be invisible to the naked eye. All this is rather awkward at first, because the object must be moved while the eye is looking through the microscope; and, in addition, if it is to be pushed to what appears to be the left-hand side of the field of view, it must actually be pulled towards the observer's right hand; and if the image is to travel up the field, that is, away from the observer as he sits at the microscope, the object must really be slipped towards him, because the lenses reverse the image. This seems a very complicated proceeding, but it soon becomes the easiest thing imaginable. At the first trial the object will be sure to leap entirely out of the field, because it will be too rapidly moved, and the motion is magnified as well as the object; but the student will become so expert that before very long he will be able to make on the stage of his microscope complicated dissections with fine needles of the internal organs of the housefly, or some other equally small insect.

The stage will probably have two springs on the upper surface, one on each side. These "spring clips" are to keep the glass slide holding the object in position, unless intentionally moved. The slide is put under the clips, and the object, provided it is itself stationary, will remain in the field, where it can be examined quietly and comfortably.

The diaphragm should always be present. It will be pierced near the edge with a series of openings of various sizes, to modify the amount of light thrown on the object, the largest opening admitting the greatest amount. The beginner will at first be disposed to use too much light; indeed this is a fault of many older microscopists. More can be seen with a moderately lighted field than when the eye is dazzled and half blinded by a fierce glare. Such a blaze is objectionable, not only because it tends to obscure the finer parts of the object, but it may lead the student or his friends to condemn the microscope as injurious to the sight—an unjust accusation more than once made. If too much light is undesirable, do not go to the opposite extreme and strain the eye by forcing it to work in semi-darkness. Keep the field sufficiently lighted to be pleasant to the sight. Turn the diaphragm until the opening giving the most agreeable effect and illuminating the object enough to show the parts clearly is under the centre of the stage opening. If the object is very thick or opaque, more light will be needed than if it were perfectly transparent; in such cases use a larger diaphragm opening.

The mirror is one of the most important parts of the stand. It should have both a concave and a plane surface, and it ought not to be less than two inches in diameter, so that it may reflect enough light and be easily handled. In the newest styles of stands the mirror is arranged to swing from side to side, so as to throw an oblique beam of light on the object, as well as to rise above the stage, so that light may be reflected down upon an opaque specimen, since it is used below the stage for the illumination of transparent substances only. This swinging arrangement is very convenient, and should be had if possible. It is, however, not absolutely necessary, as similar illumination of opaque bodies can be obtained by the "bull's eye condensing lens," a

rather expensive piece of apparatus, and somewhat difficult to manipulate successfully. But as the newest and
best stands have the swinging mirror, the condensing
lens need not be described, especially since the beginner
will not care to examine many opaque objects that will
demand stronger illumination than that of ordinary diffused daylight or common lamplight.

When ready to examine an object, the stand is placed
near the window, or, if at night, the lighted lamp is
stood near the instrument on the left-hand side and one
or two inches in front of the mirror, and the objective
is screwed on. The microscope is inclined at a convenient angle; the mirror is moved in various directions,
until the light is reflected from a white cloud, if possible, or from the lamp, onto the front of the objective,
where it can be easily seen. The eye is then placed at
the eye-piece, and if the field is but partially lighted,
as it probably will be, perhaps one-half of it being in
shadow, or only a faint trace of light visible at one side,
the mirror is slowly moved until the field is brightly
and evenly illuminated, when every part of the circular
bright space within the instrument is as well lighted as
every other part. The position of the diaphragm is
then changed, to be further altered, if necessary, after
the object has been placed on the stage. This even illumination may at first be a little troublesome to obtain,
but as in so many other actions in connection with the
microscope, a very little practice will overcome every
difficulty. The fingers are soon taught; they speed-

ily do their work without their owner's conscious bid-
ding.

The specimen to be studied may be permanently preserved, or "mounted," on a slip of glass, under a thin
cover and surrounded by Canada balsam, glycerine, or
some other preservative, thus forming preparations
called "slides," or "mounted slides," the plain piece of
glass without the object being a "slip." The addition
of the object therefore changes the slip into a slide. It
is well to remember this distinction in talking with the
dealers or sending orders by mail.

Slides can be made by the student, although to do the
work neatly and well demands some skill and considerable preliminary study of the object before it can be prepared for the mounting processes; or the slides may be
purchased. It is much better and, in the end, more satisfactory to the owner of the slides to prepare them
himself. Certain rare objects, if desired, must be bought
already mounted, but any small object naturally dry
can be so easily mounted by placing it in a drop of Canada balsam from the druggist's, and covered by a cover
of thin glass from the optician's, that for the beginner
to spend his money for "the foot of a fly," "dust from
a butterfly's wing," "the sting of a bee," or similar
slides crowding the dealers' lists and drawers, is non-
sense, unless he lives alone in the wilderness, and is ignorant of the appearance of a slide; in such a case, to
buy the mounted foot of a fly may be useful to show
what is to be aimed at in the preparation of ordinary

objects. A few properly mounted slides, however, usually accompany the stand as specimens, or the dealer
will supply them if asked. It is better to do than to
buy, and so much has been written on the subject of
microscopic mounting, and indeed all advanced workers
with the microscope are such "good fellows," they are
always so generous in giving away for the asking information that has cost much time and labor to obtain, that
the young student need never despair, nor be at a loss
as to where to go for help, if he possesses the name and
address of some microscopist and a postage-stamp or
two. Cheap little hand-books on the subject are accessible, microscopists are numerous and willing, so why
should the beginner ever be discouraged? and why should
he buy what he can make? It always adds a zest to
this work if the worker can make his own tools, and especially if he can prepare his own objects. Almost every tool needed at the beginning can be made at home.
Slides must be made at home if one desires to examine
any of the endless variety of the invisible animal and
vegetable life with which the great world teems. All
the objects referred to in this book can be studied when
only temporarily mounted; indeed, no method of pre-
serving some of them has yet been discovered or invent-
ed. They must, therefore, be studied alive or not at
all. And for the beginner this is not only the easiest,
but it is the most inspiring way.

Some things can be examined when dry. Such an
object is simply laid on a slip, placed under the spring

clips, and the low-power objective used. The ripe seeds
of wild plants are easily studied in this way, and some
of them are marvellously beautiful. Small insects can
also be looked at when dry, but the result is not always
entirely satisfactory unless they are viewed as opaque
objects. Usually most objects appear better and show
more of their structure if examined under a disk of thin
glass and surrounded by water. But seeds, scales from
butterfly's wings, and many other things, can be viewed
and preserved in a dry state by enclosing them in a cell
with a thin glass cover fastened above. This "cell"
and "cover" and fastening process will be described
presently.

All plants and animals living in water must be examined in water. To dry them and expect to learn
anything about them, or even to obtain a correct idea
of their true appearance, is a waste of time, and worse.
When your wet specimens get dry on the slide, and you
think you are seeing some wonderful things, add a drop
of water, and save yourself a probable blunder. Certain objects, naturally dry, will look better and will reveal their secrets sooner if examined wet. This is due
to optical reasons not necessary to explain here. The
observer, if he is seeking information, and not merely
pretty things to please the eye and the æsthetic fancy,
will do well if he examines naturally dry objects both
in and out of water; but things naturally wet must
never be seriously studied in a dry condition.

The most convenient size for slips is three inches in

length by one in width. Some microscopists use and recommend them two and one-half inches long by one-half an inch wide, and this will probably be the size of the slides accompanying the student's stand. They are, however, much too small; it will be better for the beginner to at once select the standard, three inches by one inch, size. These can be bought, and the writer would advise that they should be, as the edges will then be ground smooth and perhaps polished, although the last is not necessary. Slips can be cheaply cut by any glass-dealer who has a diamond or glass-cutting wheel, and if thus made, the best, whitest, smoothest, and thinnest glass should be selected. The rough edges of the home-made slips, however, are not pleasant to handle, the student who uses them taking the risk of cut fingers. Otherwise, unless they have a green color, they are as useful as the more expensive ones sold by the dealers.

A drop of water on a slip of smooth glass is not easily kept in position. When the slide is placed on the stage, and the microscope is inclined for use, the water will surely run away, and probably carry the object with it. If the microscope is not inclined, the convex surface of the drop, and its tremulous movements, will so affect the light that the image will be distorted, and the observer will obtain erroneous impressions. A piece of glass placed over the water will flatten the surface, the distortion of the image will be partially counteracted, and capillary attraction will keep the liquid from entirely running away. But ordinary glass is too thick for this purpose, consequently thin glass prepared for microscopical use must be purchased. This varies in thickness from No. 1, measuring about $\frac{1}{150}$ to $\frac{1}{200}$ inch or thinner; No. 2, about $\frac{1}{100}$; and No. 3, from $\frac{1}{50}$ to $\frac{1}{75}$ inch. No. 2 glass will be the proper thickness. It can be obtained either in circles of various sizes or in squares. For permanent mounts the circles are usually employed. For temporary purposes, for the examination of an object that is not to be preserved for future use, or when many examinations of separated parts of the same large specimen are to be made, the writer much prefers thin squares, and always uses them. They are pleasanter to handle, they are more easily wiped dry and with less liability to breakage, and their cost is somewhat less than circles of the same thickness.

The matter of cleaning thin glass is an important one, and unless the "knack" is soon learned, the beginner will be surprised at the rapidity with which his covers will disappear. This skill, however, is readily attained. The writer has had the same thin square of No. 1 glass in use for three months continuously, frequently removing and reapplying it during the five or six hours of daily evening work in which it did important service, and in the end he became quite attached to it as to a good friend. But a hasty move while cleaning it, or a little undue pressure, finally sent it on the way that thin covers often travel. To clean without much risk of breaking, take the square with two opposite edges, that is, with the edges where the glass was cut, between the thumb and finger of the left hand, and with a piece of soft, old muslin held smoothly over the thumb and forefinger of the right hand, gently wipe both surfaces at once, rotating the square when necessary. The secret of success is care, gentleness, and no wrinkles. It was probably a wrinkle in the muslin that ruined my three months' old pet cover. But a punishment is a good thing sometimes; the microscopist who should begin to think that he was skilful enough to avoid breakage of covers for more than three months, might become insufferably conceited and a nuisance to his friends.

But a glass square, however thin, dropped on a delicate animal or plant will often crush it, and destroy all resemblance to anything that ever lived. Some means must be devised for supporting it at a very short distance above the slip, so that the living creatures may have room to move about, and the plants may not be too much flattened. This is done by making a ring of cement on the slip, and thus enclosing a circular space called a *cell*, which can be made of any depth by applying more cement after the first application has dried, or by using the cement very thickly.

The opticians offer several kinds of cement for sale, all of which are useful for special purposes; but the one that seems most convenient, and one that can be easily prepared by the beginner, is simply shellac dissolved in alcohol. The solution can be made as thick as is desired by allowing some of the alcohol to evaporate, or it can be thinned by the addition of more. It should be thick enough to flow freely from a small camel's-hair brush, but not so thin as to spread in an irregular film over the glass. As shellac dissolves slowly in alcohol, it is better to add more of the latter than will be needed, and to thicken the solution by evaporation. It will keep for any length of time in a tightly closed bottle.

A ring can be built up with a camel's-hair brush, and this cement, either by the hand alone, or by a little machine called a "turn-table," manufactured for the purpose. These turn-tables are as nice and neat and beautiful as can be imagined, and they cost—the cheapest that I can find in the catalogues costs $2.50. They spin perfect circles exactly in the centre of the slip, and the result is very pretty and very desirable if the beginner can afford one, but he can get along right well without. If you have none, draw in the centre of a strip of white pasteboard the size of a slip, a circle in black ink, and use it as a guide to the brush with which you make the ring after the slip is laid on the pasteboard. Of course the hand cannot be as steady as a flat disk rapidly rotating on a central pivot, and the circles will not be as perfect, but they will be practically as useful. To get the inked circle in the centre of the paper, draw a lead-pencil line diagonally across it from each upper corner to the opposite lower one, and use the point at which the two lines cross each other as the centre of the circle. The glass slip can be kept in better position, and the whole can be turned about, if the pasteboard is fastened to a strip of wood, and a small pin is driven into each corner. When the ring is made, put the slip in a warm place until the cement is hard, or hold it over the lamp flame for a few moments at a time, taking care not to allow the shellac to boil, or the bubbles will never disappear and the ring will be weakened. These lamp-dried rings are hard as soon as cold, and they adhere so firmly that they can only be scraped off with a knife and hard work. They have the further advantage of being rapidly made.

A deeper and perhaps a somewhat neater cell can be formed from paper. Cut a circular disk, of the diameter of the ring required, from porous paper as thick as the depth of the desired cell, and from the centre cut out a smaller disk, leaving a ring with a narrow rim. Soak this ring in thin shellac cement until its pores are filled with the liquid, and hang it on a pin in a warm place to dry. Several can be prepared at once, and can be of different sizes and thickness. They are fastened to the slip by touching one side with a little shellac and pressing the glass on it and allowing it to dry, or by gently heating the slip and ring over the lamp. It is a good thing to prepare several slips at one time, so as to have them ready for an emergency, as, for instance, after an excellent gathering of microscopical material has been made, and the student is so anxious to see what he has that he cannot take time to clean the slide and cover after a hasty glance for rarities, but must have another ready at a moment's notice.

To permanently mount dry objects, such as pollen, seeds, scales from insect wings, and other things suitable for this method of preservation, arrange the specimen in the cell, place the cover over it—preferably a circle in this case, the diameter of the cell being a little greater than that of the cover—so that the cement shall project a short distance beyond the edges of the thin glass, and with a camel's-hair brush paint a thin layer of shellac over the place where the cover and ring meet. There should be but little cement on the brush for the first coat, because if too much is used, or it is too thin, it will probably run into the cell by capillary attraction and spoil the object. This is one great trouble in all microscopical mounting. But after the first coat is dry, another is to be added, and repeated until the cover is firmly fastened to the ring. "Brown's Rubber Cement," for sale by the dealers, is useful for this purpose, as it is very fluid, dries with great rapidity, and has little tendency to "run under."

The cell having been made, the object is to be placed within it in a drop of water, the thin cover dropped over it, and the preparation will then be ready for examination. But how is this minute, generally invisible object to be got into the cell? A glass tube about one-tenth inch in inside diameter, and as long as may be convenient, several needles in wooden handles, and a camel's-hair brush, with a small smooth stick thrust into the quill, will be needed.

The needles are used for spreading any small mass

evenly over the cell, and in disentangling and arranging the parts of any comparatively large object, as well as for lifting the thin cover from the cell so that it can be easily seized by the fingers, or for tilting it up in the box, where the thin squares should always be kept. Fresh-water Algæ (Chapter III.), for instance, found so abundantly in almost all still water, where they often form delicate green clouds, or thread-like streamers adhering to other plants, dead leaves, or waterlogged sticks, are almost sure to be transferred to the slip in a heaped up and tangled mass, which only two needles with gently persuasive movements can straighten out for microscopic study. If an attempt is made to examine such a confused heap, the thin cover cannot be forced to lie flat without crushing the delicate specimens, and if the cover is tilted the objective cannot be properly focussed. To make these useful tools, with pliers thrust fine needles head first into parlor-matches, after the phosphorous ends have been cut off. These round sticks make handles convenient in length and pleasant to use. It is well to have half a dozen or more of these needle-bearing matches lying where they can be picked up whenever wanted. If the student desires to dissect insects, nothing can be so useful for cutting and tearing minute parts and for separating delicate tissues or organs as fine needles. No knives have been made to equal them for this purpose.

The glass tube is the "dipping-tube." It is really one of the most important little pieces of apparatus that the microscopist can have on his table, if he intends to study aquatic life. With it he can pick up any small object that may be visible in the water, transfer any selected matters to the slip, or make the dip that is made by faith, with the assurance that although the tube may seem to be filled with water only, it will be pretty sure to have captured something interesting, novel, or beautiful. He can fill the tube with water, and allow it to escape in a miniature torrent, or drop by drop, or he can allow a drop to enter and a drop to flow slowly out at his will. Some workers prefer a tube with a hollow rubber bulb attached, by which the water and contained objects are drawn up by the expanding ball, and forced out by its compression. The writer is prejudiced in favor of the simple tube, as it is less complicated, more easily cleansed, and its contents are more completely under control. To use it, place the tip of the forefinger firmly over one end, and dip the other into the water above and near to the object desired; lift up the finger, and the water will rush in until it is level with that on the outside; close the upper end again, remove the tube, and the water will remain in it as long as the finger stops the upper opening; remove the finger and the water will at once flow out. By the proper regulation of the pressure and movements of the finger, the water can be made to escape drop by drop or in a sudden rush. In this way any small aquatic object can be easily transferred to the slip, and as readily washed off by a sudden outward flow from a full tube.

3

Until recently I supposed this little affair was common property, and that the principle on which it acts was understood by everybody. But when I called on a gentleman, a member of a scientific society, to obtain some water in which certain plants were growing, he expressed surprise at the performance, and called his wife to witness a new and curious method of taking up water with nothing but a glass tube and a finger. His astonishment was amusing; but how much more so was that of a druggist who had a teaspoonful of deposit at the bottom of a conical glass vessel with a quart of water above it, and who, after running about for bottles and jars to hold this water, which he thought must be poured off, returned to find the deposit removed, and in a small phial in my pocket, the quart of water remaining undisturbed. "Why," he said, "that is strange. I never saw the like before. How did you do it?"

It is often convenient to have several dipping-tubes, some straight, others drawn out to a point, and some curved so as to be readily directed into a narrow corner. A glass tube is easily pulled out to a fine extremity, or variously curved when softened in an alcohol flame. But a spirit-lamp may not always be within reach, and is not necessary, for the student can make a Bunsen burner almost without cost, and use it successfully if his home is supplied with illuminating gas. Prof. Austin C. Apgar, in *Science News and Boston Journal of Chemistry*, has, under the title "A Bunsen burner for two cents," recently described a simple piece of appara-

tus that is a boon to any one desiring to do a little ama-
teur glass-blowing. A strip of tin about six inches long
and two wide is rolled, without solder or fastening of
any kind, into a tube about half an inch in diameter,
after two holes, each about one-fourth inch in diameter,
have been punched so that they shall be on opposite
sides of the tube, and high enough to be a short distance above the tip of the gas-burner. This simple arrangement is forced over an ordinary burner, so that the
holes are just above the tip, the spring of the tin holding it in place; the gas is lighted at the upper end,
where it burns without smoke and gives a strong heat,
the flame being easily regulated, and, with ordinary care,
not flashing into the tube. It is entirely successful.

Evaporation of the water will take place from beneath the thin cover, sometimes quite rapidly, and the
observer will at first be surprised at the way in which
his objects will be swept out of the field before an ad-
vancing wave that leaves the glass nearly dry behind it.
The water in the cell is drying up, and a fresh supply
must be added if the objects are not to be entirely lost.
Here is another advantage in using square covers on circular cells. The four corners project beyond the cement
ring, and by applying the camel's-hair brush, wet with
water, to the slide beneath any one of these projections,
the drop will run in and fill the cell by capillary attraction. This supply is much more easily added than if
circular covers are used, and after a little experience the
fresh drops can be applied while the eye is at the eye-

piece, the hand alone guiding the wet brush, and the eye taking note of the rush of the incoming wave and the result. The student will soon become such an adept that he will be able to add so small a supply at each touch of the wet brush that the movement of the capillary wave will not be strong enough to float the object out of the field.

But it often happens that a certain specimen is to be studied for a long time, a whole evening, for instance, and to be continually supplying the loss by evaporation is not convenient—the student often becoming so absorbed that he forgets this one of nature's laws until he suffers the penalty, and probably loses his object. At such a time an arrangement is wanted for supplying fresh water continuously and without demanding much attention, and such a contrivance is easily made. With a triangular file cut one of the smallest homœopathic phials in two, throw away the upper half, and cement the lower to a little oblong or square piece of ordinary glass or broken slip. Attach this to the slide by a drop of glycerine, taking care not to use too much, or the square will glide out of place when inclined. Fill the bottle with water, coil into it one end of a doubled, loosely twisted thread of sewing-cotton, and place the other end in contact with one side of the cover, as shown in Fig. 3. The water will pass down the thread to one edge of the cell, where it will flow under as it evaporates from the other three sides. This usually works well, the secret of success being to have the reservoir not more than three-quarters of an inch from the cell, to keep it always full of water, and to have the doubled thread applied closely against the cover. If the water supply is too great, and the cell is disposed to overflow, shorten the end of the thread against

Fig. 3.—A Growing-slide.

the cover; if not enough, lengthen it, and do not allow the thread to touch the slide in its course from the reservoir to the cell.

Again, the observer frequently wants to make a growing-cell of the slide on which he may accidentally have placed a desirable or beautiful object; that is, he desires to preserve the specimen for several days in the cell without disturbing it, and so taking the risk of losing the invisible thing. He may also wish to watch its growth and development. A reservoir for a water supply is necessary; an "individual" butter-dish makes a good one. Place the slide across the dish, apply a doubled thread of sewing-cotton along one side of the square cover, so that each end shall hang down into the dish, and fill the latter with water, which will then pass up and along the thread, and keep the cell full for as long as may be desired. The only objection to this little affair is that, after a few days' use, the salts in the water will crystallize on the cover, and so cut off part of the oxygen supply. But no growing-cell is free from some objectionable features; none can quite imitate the natural conditions, and the animal or plant dies before long, either falling to pieces or becoming buried beneath a mass of fungi. This one will supply an abundance of water, if the water in the dish is always kept in contact with the lower surface of the slide. This, and the absolute contact of the thread with the edge of the cover, are the only things whose absence will result in defeat.

As the reader already understands, the object must never be examined in water without being covered by either a thin glass circle or square; the importance of this little piece of glass must not be forgotten. But very often, in lowering it over the wet specimen, small bubbles of air will be caught and not noticed until magnified, when, if seen for the first time, they appear wonderful, if not startling. Some strange statements have been made. and discoveries announced whose only foundation has been minute air-bubbles that the observer did not recognize. A man once described a marvellous something that he had found in a cancer, which turned out to be a magnified air-bubble. These little air-drops always play an amusing part at the beginning of the microscopist's career. In Fig. 4 are shown several of different sizes. Let the student examine a drop of saliva or of soapsuds, and he will in future be able to recognize the troublesome things. Pictures or words cannot convey so true an idea of their appearance as a

Fig. 4.—Air-bubbles.

single sight of the bubbles themselves. At times they
become entangled in the parts of an object in such numbers as to interfere with its examination. In these cases
nothing can be done except to lift the cover on the
point of the needle, and slowly lower it, or remove it
entirely, add more water, and reapply it carefully. In
appearance the bubbles are usually circular, with a broad
black border which varies in width and depth of color
as the objective is raised or lowered. Near the margin
is a bright ring, and in the centre a bright spot. They
often float about, and this movement adds much to the
wonder with which the beginner usually regards them.

If the student will have a note-book in which to jot
down his observations, or to keep a list of the objects
examined, it will not only aid him in forming habits of
accurate observation, but will be of great interest when
he has become an accomplished microscopist. The en-
try may be very simple, and may be made to serve as a
memorandum of items to refresh the memory. Here is
an example from a boy's note-book: "June 15, 1884—
Came across a pool near the toll-gate with the water
colored green, and found the color was caused by a great
quantity of Volvox—small green globes rolling about in
the water. Volvox is said to be a plant. Wonder if it
is. What are the darker balls inside of some of them?"
He answered all these queries later in his experience.

If you can draw the microscopic objects that interest
you most, although the sketches may not be quite artis-
tic they will help you to remember, and a collection

of such drawings will be as interesting and valuable as
the note-book.   In talking to friends about microscopic
matters, a single rough drawing will do more to help
them understand than many words.   And if you can
look at the object and make the sketch, you will like it
better and do yourself more good than if you bought
and used the drawing apparatus called a camera lucida,
for sale by the dealers.   This camera lucida is a glass
prism, so arranged that when it is put over the eye-piece,
and the microscope is placed in a horizontal or inclined
position, the magnified image seems to be reflected down
on a sheet of paper spread on the table just under the
camera, but of course with a space of several inches between them.   By placing the eye in the proper position,
and looking down towards the table through the edge of
the prism, the image and the pencil-point can both be
seen at once and the outlines traced.   It is a rather expensive apparatus, and difficult to use without a good
deal of practice, but if you want a
simple arrangement that you can
make, try the one shown in Fig. 5.

From a piece of thin sheet brass
or tin, cut with scissors a strip half
an inch wide and long enough for
one end to pass around the upper
part of the eye-piece, and the other
to be bent into a handle like a small
hollow square.   Cut another strip about one inch long
and one-fourth wide, and double it lengthwise so that it

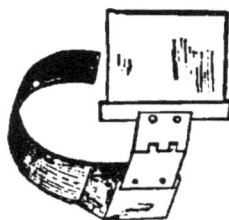

Fig. 5. — Reflector for
Drawing the Magni-
fied Object.

will still be an inch long, but one-eighth of an inch broad. Take one of the small brass hinges to be had for a cent, solder one end to the hollow handle and the other to the narrow doubled strip; into this narrow piece place a thin glass square, the thinner the better, and the instrument is done. To use it, turn the microscope horizontal, have a faint light on the object and a strong one on the paper, bend the strip of brass around the upper part of the eye-piece so it will not slip, the hollow handle and hinge being directed towards the table, and move the hinge until the thin cover is placed obliquely in front of the eye-glass of the eye-piece. Look down through the glass square towards the paper on the table, and the image of the object on the stage will seem to be thrown on the white surface, where it can be traced with a pencil. The image is really reflected from the surface of the thin square, and the pencil is seen through it, but the eye unconsciously combines them so that both are seen together. The secret of success here is a faint light on the object, a strong one on the paper, and a *thin* glass square. A long, sharp pencil-point is also an advantage.

A micrometer is for measuring objects under the microscope. It is made by ruling a number of short lines on glass, the spaces between the lines varying from $\frac{1}{100}$ to $\frac{1}{1000}$ inch or less. Micrometers are said to have been ruled with one million lines to the inch, but no human eye using the best and highest power objectives has ever seen them. All micrometers are ruled by a machine made for the purpose.

3*

The beginner will not need one, but he may desire to know how to use it. Place the micrometer on the stage, turn the microscope horizontal with the reflector referred to above, fitted to the eye-piece. With the low-power objective focus the lines that are $\frac{1}{100}$ inch apart, and draw them on the paper. Do the same with every objective, drawing the $\frac{1}{1000}$ inch spaces with the $\frac{1}{4}$ or $\frac{1}{8}$ lenses. These drawings will form the scale for measuring the drawings of the magnified objects. Thus, if the magnified object, when drawn, occupies two spaces of your paper scale made from the $\frac{1}{100}$ inch micrometer spaces, the object will measure $\frac{2}{100}$, or $\frac{1}{50}$ inch in length; if five spaces of your scale, then it will measure $\frac{5}{100}$, or $\frac{1}{20}$ inch long; if only one-half a space of your scale, then it will measure one-half of $\frac{1}{100}$ of an inch; if one-fourth of your scale space, then its actual length will be $\frac{1}{400}$ inch. If the $\frac{1}{4}$ or $\frac{1}{8}$ objective is used in making your scale from the $\frac{1}{1000}$ inch micrometer spaces, then each division on the paper will represent $\frac{1}{1000}$ inch, and if the drawing of the object measures two of these spaces on your scale, the real length of the object will be $\frac{2}{1000}$ inch, or $\frac{1}{500}$. It is perceived that the stage micrometer cannot be used for measuring objects directly, but only by applying the drawing of the magnified micrometer spaces to the drawing of the magnified object.

The micrometer can also be used to ascertain the power of the microscope. If each of the $\frac{1}{100}$ inch spaces measures, when drawn on the paper, $\frac{1}{10}$ inch, that com-

bination of eye-piece and objective will have a magnifying power of ten diameters; if each $\frac{1}{100}$ inch micrometer space measures $\frac{4}{10}$ inch, the power will be forty
diameters; each, therefore, corresponds to ten times. If
the $\frac{1}{1000}$ inch micrometer spaces measure, when drawn,
$\frac{1}{10}$ inch, then each tenth corresponds to a power of one
hundred times; therefore, if the $\frac{1}{1000}$ inch spaces, when
magnified, measure ten-tenths, the power of that eye-
piece and objective is of course one thousand diameters,
or ten times one hundred; if five-tenths, then five times
one hundred.

The owner of a microscope should never take a walk
in the country without one or two wide-mouthed bottles in his pocket. Empty morphia bottles, to be had
of any druggist, are convenient for small collections;
for greater quantities an empty quinine bottle, and for
still larger gatherings of aquatic plants the ordinary
glass fruit-jar is admirable if a string is added for a handle. No bottle should be entirely filled and corked, or
all animal life will be animal death before the microscope is reached. Leave a large space for air between
the cork and the water.

Those desiring information as to the optical construction of the compound microscope, the uses of the numerous pieces of apparatus often used for advanced work,
and about the methods of permanently mounting microscopic objects, may advantageously consult the following publications:

"How to Use the Microscope." 16mo. By John Phinn. New York. "How to See with the Microscope." 12mo. By Dr. J. E. Smith. Chicago. "How to Work with the Microscope." 8vo. By Dr. Lionel S. Beale. London. "The Microscope." Small 4to By Dr. W. B. Carpenter. London. "The Micrographic Diction- ary." 8vo. London. "Manual of Microscopic Mounting." 8vo. By John N. Martin. Philadelphia. "The Preparation and Mount- ing of Microscopic Objects." 16mo. By Thomas Davies. London. *The American Monthly Microscopical Journal.* Washington, D. C. *The Microscope: an Illustrated Monthly Journal.* Ann Arbor, Mich.

CHAPTER II.

COMMON AQUATIC PLANTS USEFUL TO THE MICROSCOPIST.

Ranunculus. —Nymphæa. —Myriophyllum. —Utricularia. —Ceratophyllum.—Lemna.—Anacharis.—Vallisneria.—Sphagnum.—Riccia.

THERE are several common plants floating freely in
the water, or more or less firmly rooted in the mud at
the bottom of shallow ponds and slowly flowing streams,
that are important to the student of microscopic aquatic
life.   This may be either through their own interesting
or peculiar structure, or on account of the minute plants
and animals living among their tangled leaves or attached to the stem and other parts, these entangled objects being, therefore, more easily and surely captured
by transferring the larger visible growths to a small
vessel of water than in any other way.   Most of these
aquatic plants have their leaves divided into fine, thread-
like leaflets.   They have " dissected leaves," as the botanist names them, and they become the favorite resorts
of invisible animals which attach themselves to the narrow divisions, and feed on the free-swimming kinds that
also find the same places attractive.   So, if the student
desires to gather microscopic material, let him find any
of the following plants and he will be quite sure to get
what he wants.   But he must remember that by lifting

them out of the water very many of the creatures he
most desires will be washed away.   The plants should
be slowly and carefully drawn to the shore, and lifted
out in a tin dipper and poured into a wide-mouthed
bottle.   The small tin dipper will prove a very convenient implement for all kinds of microscopical collecting,
as a handle of any length can be made by thrusting a
stick into the hollow handle of the dipper.   If the latter, however, is not accessible, the plants may be gently
pushed into the bottle, after it has been partly sunk so
that it lies parallel with the surface of the water.

Many of our commonest aquatic plants have no common English names, probably because most of them
bear the smallest and least showy flowers of all blooming plants, and therefore do not attract the attention of
the ordinary observer.   In referring to them, the beginner must use the scientific names, or learn the meaning
of the Latin words and use the translation, usually with
awkward results.   It sounds better and is quite as easy
to speak of *Myriophyllum* as of the "thousand-leaved
plant," which the word means.   Many plants might be
styled thousand-leaved; another common aquatic one,
for instance, which often grows in the same pond with
Myriophyllum, the *Ceratophyllum*, called "hornwort"
because the leaves are rather stiff and horny ; and *Lémna*, as a word, is prettier and more appropriate than
"duckmeat," an ugly term and meaningless, because
ducks have nothing to do with the plant.

If the reader is not already familiar with the appearance of the following forms, he need have no trouble in learning their names, although he may not have studied botany; he has only to compare the leaves with the figures in this chapter.   It is, of course, understood that there are many aquatic plants not here referred to, only those being included in this list which afford the most certain supply of microscopic life.   The leaves of many water-plants fall against the stem and cling together when lifted into the air; but if the student will place a small part of the plant in a saucer ("individual" butter dishes are good for this purpose), he can float them out against the white surface and so compare them with the figures.

RANÚNCULUS AQUÁTILIS (Fig. 6).

A part of the stem and a single leaf of this plant are shown about natural size in the figure (Fig. 6).   It is quite common in ponds and slowly flowing streams.   The leaves are dissected into fine, rather stiff and hair-like parts, to which many minute animals, such as *Rotifers* (Chapter VIII.), *Vorticellas* (Chapter V.), and *Stentors* (Chapter V.) are fond of attaching themselves.   The leaves are placed above each other on opposite sides of

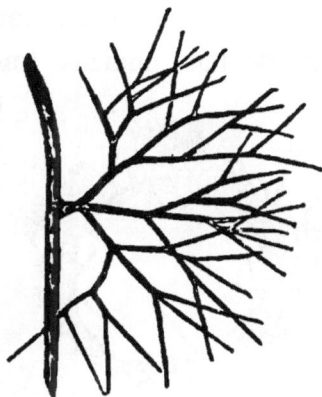

Fig. 6.—Leaf of Ranúnculus aquátilis.

the rather brittle stem, and usually quite wide apart.

The whole plant is under water except at flowering-time, when it raises a delicate stalk above the surface, and blooms with a single white flower closely resembling the common yellow "buttercup" of the fields.

NYMPHÆA ODORÁTA (White Water-lily, Fig. 7).

Every one is familiar with this beautiful flower, that "marvel of bloom and grace," and the large, almost circular, floating leaves. It is to the under-surface of the latter that the microscopist often goes for several forms of case-building Rotifers, with the certainty of always finding them, together with many and various kinds of minute animal life. It is also an excellent place to search for worms. You will usually find these creatures if the surface is gently scraped and the dark mass obtained is examined in water.

But if the scented blossom is beautiful to the ordinary observer, the interior of the flower-stems and leaf-stalks has charms known only to the microscopist.

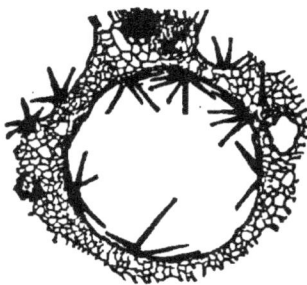

Fig. 7.—Peduncle of Nymphæa odoráta; transverse section.

Cut a thin slice from either of those parts and examine it. The sides of the wide openings made by cutting across the internal tubes are studded with crystalline stars (Fig. 7). Three-pointed, four and five pointed, they sparkle there like diamonds, yet they were formed in darkness, and in darkness act their part in the life of the plant. What that part is we can only guess. Botanists call them internal hairs; but they are hard, sharp, and brittle. They are hollow, too, and their surface is roughened by minute elevations, as though fairy fingers had sprinkled them with crystal grains. I never see a white water-lily without in imagination seeing those long stalks rising out of the black mud up through the dark water, with their entire length illuminated by the sparkling of these internal star-like gems. The whole plant contains them, even the root. The common "spatter-dock"—hideous name!—the *Núphar*, also conceals similar stellate hairs within its stems, but they are there larger and coarser, as becomes a coarser plant. The leaves of the Núphar, however, are not a good microscopical hunting-ground, as they usually stand high above the water.

MYRIOPHÝLLUM (Fig. 8).

This is not rare in shallow ponds and slow streams; it even occurs in running water, but there it is not worth gathering, so far as any adherent microscopical life is concerned. Indeed, no running water is a good locality for free-swimming creatures, because the current sweeps them away, and so scatters them that it is not possible to make a collection. But where Myriophýllum grows it usually grows abundantly. It forms long green streamers, round and thick, sometimes more than an inch in diameter and several feet long, yet it looks soft and feathery. The leaves are very numerous, and each

set is arranged in a circle around the stem; they are in
"whorls," as the botanist calls the arrangement. One
such whorl is shown in Fig. 8. Five dissected leaves
are there drawn, but whorls
sometimes occur with three
or four, the number helping
to distinguish the species,
of which there are several.
They all resemble one another when in the water. The
parts of the leaf are fine,
soft, and hair-like, those nearest the stem of the plant
being the longest. They are
very numerous and close to-
gether, thus giving the floating streamers their peculiar thick and soft appearance, and making them an excellent place for the microscopist to explore.

Fig. 8.—Whorl of Myriophýllum
Leaves.

To compare with Fig. 8 a feathery plant which the
collector does not know, select a circle of leaves, cut the
stem close above and below it, and after floating the
separated whorl in a saucer as already directed, or spreading it out on white paper, compare its leaves with those
figured. The leaves vary in size in different parts of
the plant, the uppermost being smallest and youngest,
the lower the oldest and largest.

There is another rather common aquatic plant called
Proserpináca, or "mermaid-weed," which so closely resembles Myriophyllum when in the water that it has

often been mistaken for it.   To make such an error is of no great consequence, unless it should lead the observer to imagine, as it once did the writer, that he has found a rare species of Myriophyllum.   Yet it is always pleasant, if nothing else, to feel sure, and it is more than pleasant to have a reputation for accurate observation. Proserpinaca, however, is as useful a trap as Myriophyllum, and it can be easily distinguished because the dissected leaves are not in an exact circle around the stem : one is on one side, the next a little further round and a little higher on the stem, another still further round and nearer the first, but still higher.   They are what the botanist calls alternate.

Either of these plants is a specially good place for attached diatoms (Chapter III.).

UTRICULÁRIA (Figs. 9 and 10).

Of all our water-plants with finely divided leaves, Utriculária is probably the most interesting in itself, and one that can always be recog-
nized at a glance.   It is found in long, somewhat branching streamers, floating freely below the surface or very slightly rooted.   A leaf of *Utriculária vulgáris*, a common species, is shown somewhat enlarged in Fig. 9, with the peculiar hollow bladders, or "utricles," that distinguish it from all other plants, and give it one of its

Fig. 9.—A Leaf of Utriculária.

scientific names. These utricles are almost always conspicuous when the plant is taken from the water, as small, green, semi-transparent particles attached to the leaves. They are not unlike small pieces of jelly in appearance, until examined with the microscope, when their remarkable structure is seen. Until within a few years they were supposed to act as air-sacs to help the plant float. It was even said that they became filled with air or gas at flowering-time, and so lifted the flower-stalk and the bloom above the water. This was interesting, but the truth is more interesting and startling. The plant actually feeds on animals. These bladder-like bodies are the food-traps, the mouths and the stomachs of the Utricularia.

Under the microscope they are seen to be hollow, oval bodies, with a narrow, almost straight anterior end, and several long bristles projecting forward or away from the utricle, these bristles probably serving as a guide to an opening at their base. The little animal swims or crawls against a bristle, and naturally moves down it towards the opening in the utricle, which it finds closed by a transparent colorless curtain; this it pushes aside and passes on into the trap. The curtain-like valve is attached by its upper and lateral margins, therefore hanging before the opening in the utricle, and swinging inward, but so arranged that it cannot be forced outward by any creature small enough to pass within. Indeed, the power that the valve seems to exert is somewhat astonishing. Small fish have been found with the tail or even the head inside the utricle, and firmly held by the pressure of the valve.   In these cases, however, it seems probable that the struggles of the dying fish may have wedged it fast, rather than that the valve has held it.   Small worms and worm-like larvæ have been found half in and half out of these fatal traps, for once past the curtain-like valve the little animal never escapes.   And no sooner has it entered than it begins to show signs of discomfort; if it has a shell it withdraws its legs and head and closes the shell; if a worm or animalcule it speedily becomes languid, its movements cease, and it finally dies, as does every creature that ventures into Utricularia's utricles, which evidently contain something more than simple water.   If these bladders are torn to pieces under the microscope with the needles, the remains of many kinds of minute creatures will be seen, the soft parts of the captives having been dissolved and absorbed, and gone to nourish the plant.   The whole inner surface of the utricles is lined by innumerable colorless four-parted bodies, one of which is shown

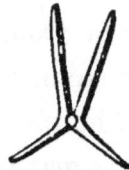

Fig. 10. — Quadri-fid Process from Inner Surface of Utricle of Utricularia.

much magnified in Fig. 10.   They are distinctly visible only when the utricle is torn to pieces.   They are said to absorb the fluid in which the entrapped animals are dissolved.

CERATOPHÝLLUM DÉMERSUM (Fig. 11).

This is commoner and more abundant than Myrio-phyllum, for which it is often mistaken, although the

two have only a remote general likeness. The leaves of
Myriophyllum are fine and soft, those of Ceratophyllum
rather coarse and stiff. In the latter they are whorled
with six to eight in each circle, but instead of being divided on each side down to the middle line (the midrib),

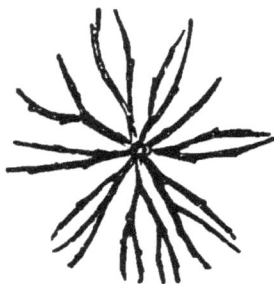

as in Myriophyllum, they appear
to separate into two narrow parts
near the stem, while each division
then often divides into two other
parts. Both these arrangements are
represented in Fig. 11, where the
whorl is shown separated, as was
done in Myriophyllum. The leaves
always bear several very small but

Fig. 11.—Whorl of Leaves of
Ceratophyllum.

visible spines on their sides, as in the figure, and when
taken from the water they usually do not fall against
the stem.

The plant is found in still, shallow places, growing in
thick masses and often considerably branched. It makes
an excellent retreat for certain Rotifers and worms, but
the leaves are so heavy and stiff that they are not as
easily prepared for microscopical examination as are
those of Myriophyllum; they often refuse to lie flat,
and thus tilt the cover glass and allow the water to run
away. But with neither of these plants will the student
try to place an entire whorl in the cell. It is always
best to clip off with scissors a part of a single leaf, and
examine it for whatever may be attached. Work with
the microscope is delicate work, and the smaller the ob-

ject, within certain limits, the better. Many beginners make the mistake of trying to examine too large a specimen or too much of a mass at once.

LÉMNA POLYRRHÍZA (Fig. 12) AND LÉMNA MÍNOR (Fig. 13).· DUCKMEAT.

These are small plants, very common, and often so abundant that the entire surface of large ponds is covered by them as by a green carpet. The water in such cases is so completely covered and concealed that the observer is for a moment tempted to step on it. The above two species resemble each other, yet they differ so widely that a glance will distinguish them. Each consists of a small green, more or less oval leaf or frond floating on the water, with one or more rootlets hanging from beneath, but never taking root in the mud. Usually two, three, or four fronds are attached together, so as to form an irregular star. *Lémna polyrrhiza*, the many-rooted one (Fig. 12), has the largest fronds, is a deeper green, and, as its specific name signifies, has many rootlets, often a dozen, hanging in a cluster from each.

Fig. 12.—Lémna polyrrhiza.

It can always be known by this root-cluster and by the dull purple color of the lower surface. It seems to like the sun better than *Lémna mínor*, and is oftener found abundantly on open ponds, while the latter appears to prefer ditches with high banks and shade.

Lemna mínor (Fig. 13) has smaller, more oval and

thinner fronds. It is lighter green in color, the lower surface is never purplish, and it has but one rootlet to each frond. Both species have a curious little cap on the free end of each rootlet. It is more easily seen with the naked eye on Lemna polyrrhíza, where it is usually darker than the rest of the rootlet.

Fig. 13. — Lémua Minor.

There are several other species, but they are so seldom found that they are not included in this list. They all multiply by the growth of young fronds from the edges of the old and mature. This accounts for the clusters so commonly seen. They also bloom, but the flowers are extremely small and are rarely observed. The student will be fortunate to find specimens in blossom. The flowers burst out of the margin of the frond, and consist of only those parts needed to fertilize and mature the few small seeds.

The rootlets are valuable to the microscopist, as they are favorite places for many just such creatures as he most wants. The lower surface of the fronds, especially of Lemna polyrrhíza, should be gently scraped in a drop of water for Rotifers not often found elsewhere. It is also much visited by small worms, but not so frequently as the leaves of the white water-lily.

ANÁCHARIS CANADÉNSIS (Fig. 14).

This is readily recognized by the arrangement of the leaves in circles, or whorls, of three each, two of which

are shown in Fig. 14. The stem is brittle, and fragments easily take root, so that the plant spreads rapidly.
Having been accidentally introduced into
England, it is said to have grown so fast
that it has choked up some of the shallow-
er streams and to have become a nuisance.
It is abundant in this its native country,
but it never acts so badly here. The whole
plant is semitransparent, with leaves about

Fig. 14. — Aná-
charis Cana-
densis.

half an inch long springing directly from the stem, and
tapering to the point. These leaves, under the microscope, exhibit a remarkable phenomenon.

All plants are formed of cells, or cavities of various
sizes and shapes, surrounded on all sides by a delicate
membrane called the cell wall. The cells are seldom
empty. Their contents are chiefly the soft, colorless,
jelly-like substance called vegetable protoplasm, and the
small green grains (the chlorophyl) which give the
green color to the plant. In *Anácharis* the walls of
the leaf-cells are transparent, so that the microscope
shows a part of what is taking place within the cell; and
it is a wonderful sight, for the protoplasm is slowly
moving around the walls, carrying the chlorophyl
grains with it. Up one side of that microscopic cell
travels the strange procession; across, down, and up,
slowly and steadily the stream and the grains move
round and round. Sometimes a little thread of colorless
protoplasm leaves the main current and starts across by
a shorter road, and sometimes the current pauses, stops,

4

and refuses to move again. The streams in two cells lying side by side may flow in the same or in opposite directions, with only the thin wall between them. What causes these remarkable movements is not known. Cold seems to retard, and warmth to hasten the flow, and often, when the chlorophyl has increased so that the green grains crowd the cells, the circulation ceases, apparently because the chlorophyl has not enough space for free movement. The botanists call this circulation of the protoplasm cyclósis. It is also finely seen in the long, narrow, ribbon-like leaves of *Vallisneria*, an abundant and common plant in slowly flowing streams.

To show the cyclósis, the Anacharis leaf needs only to be cut close to the stem, placed in the cell in water, covered by a thin glass, and examined by a high-power objective. The one-inch glass will not show it.

The plant is a fruitful source of supply for our two common species of Hydra (Chapter VI.), which often occur there so plentifully that two or three hang from almost every leaf.

SPHAGNUM MOSS (Fig. 15).

On the wet shores of shady bogs this pale-green moss grows in great patches, thick, soft, and elastic. It is a beautiful plant anywhere, but it is especially so when it appears greenly glimmering beneath the shallow water, while the shadows of elder and azalea, and the broad leaves of the tangled smilax vines, make the neighboring thicket dim and cool, even when the hot sun smites the bordering fields. In such pleasant surroundings Rhizopods (Chapter IV.) and Infusoria (Chapter V.) are found in abundance. For the former it is an unfailing source of supply. The water pressed out of a little pinch of the moss will be sure to contain many individuals and species. From a single small bunch Dr. Leidy, when studying the Rhizopods, obtained thirty-eight species and many individuals of those animals, besides numerous active diatoms (Chapter III.) and desmids (Chapter III.).

The leaves make exquisite microscopic objects, on account of their curious and beautiful structure. Each leaf is formed of two kinds of cells, *a* and *b* (Fig. 15). The large ones, *a*, will, when magnified, immediately attract the attention. They are hollow, and usually empty, and they have a spiral thread running around the walls. At certain stages of growth the cell-wall also has one or more small openings, *c*, so that the water is able to pass in and fill the cell. This may explain why the plant retains moisture for so long, and why it is so easily wetted.

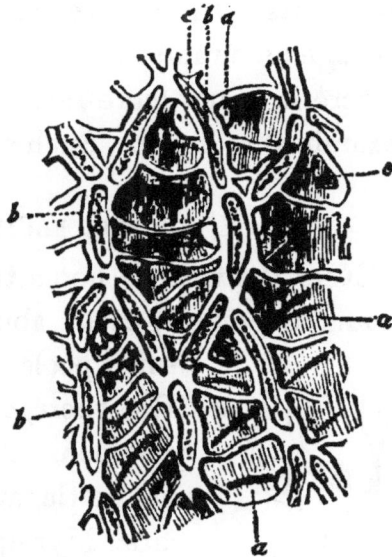

Fig. 15.—Portion of Leaf of Sphagnum.

The second kind of cells, *b*, are found between the

large ones. They are much smaller, narrower, and commonly contain chlorophyl grains, which, while usually not abundant enough to tint the whole moss a bright green, yet give it that beautiful pale hue almost characteristic of it. These cells will probably need to be searched for the first time the beginner studies a sphagnum leaf, as they are not apt to catch the eye.

The moss seems to have no roots. The lowest parts of the thick mass which it makes are usually dark and partially decayed, and it is there that the Rhizopods are most abundantly found, although many sun-loving forms are equally numerous in the brighter, better lighted upper parts. On no account should the student pass a sphagnum swamp, nor even a little patch in those places where it grows more rarely, without taking some to be examined at home. Such a gathering will always pay.

RÍCCIA FLÚITANS (Fig. 16).

Near the writer's home this little floating plant (pronounced *ricksia*) is so abundant that it often covers

Fig. 16.—Riccia flú-
itans.

small pools with a layer two inches deep. Elsewhere, on larger ponds, it is not uncommon. It often comes to the collecting-bottle tangled in the leaves of Utricularia, Myriophyllum, or Ceratophyllum, or it floats on still waters in little patches like islands. Its form is seen in Fig. 16. It has no leaves, indeed it is all leaf; the botanist calls it a radiately expanding frond, with narrow divisions, whose ends are notched. The plant is green, and may be an inch or more wide when spread out. It is often larger and more branched than shown in the figure. It has no roots, but floats freely wherever the currents or the winds send it. Shady places seem its favorite haunts.

As a microscopic object it is rather large and thick, but it forms a good place to examine for certain Algæ (Chapter III.) which tangle themselves about it in fine green threads, appear to favor it, and may often be seen with the naked eye if the single frond is placed in water above a white surface.

CHAPTER III.

DESMIDS, DIATOMS, AND FRESH-WATER ALGÆ.

THE desmids and diatoms are two closely related groups of minute aquatic plants which the beginner will at first probably have some trouble to distinguish from each other; yet after a very little experience he will be able to recognize them at a glance. Both are plants formed of only a single cell, but their beauty and variety of form, their peculiar movements and wonderful structure, place them among the most attractive of microscopic objects. And they are among the most frequent. Scarcely a drop of water from a pool in spring or summer can be examined without showing a desmid or a diatom.

The desmids are usually found in the freshest and sweetest water. In salt or brackish marshes, where diatoms flourish as well as in a mill-pond, desmids never occur. They also seem to prefer open pools on which the sun is brightest and the shadows fewest, where they probably seek warmth rather than the strong light, for they seldom form patches on the mud as the diatoms do, but adhere to the stems of other plants in a green film, or conceal themselves among the dissected leaves of the aquatic vegetation, or among tangled masses of Algæ.

A living desmid is always green; a living diatom is always brown. This difference in color makes it easy to distinguish the two groups of plants, but there are other points that can be used by even a color-blind student. The cell-wall of the desmid—that is, the thin sack which surrounds the soft green contents—is soft and flexible. If the cover-glass is pressed down firmly with a needle the desmid can be flattened or squeezed out of shape, and the cell-wall can often be broken, so that the green and colorless mixture of jelly-like matter filling the plant is forced out. The cell-wall of a diatom is hard and brittle. The cover-glass may be pressed upon until the glass breaks, yet the diatom will not be flattened nor its shape changed. It may roll over and look quite different in form when viewed in another position, but it will probably roll back and appear as at first. It can be broken, however; and it does so as if made of glass or some other hard and brittle material, and the yellowish-brown contents, may flow out, but the broken place will not be a hole with irregular edges, as it was in the crushed desmid; the edges will be sharp and angular, and the diatom will probably break into several fragments. Yet with the most skilful manipulation it is rather difficult to purposely break any but the largest of the diatoms, few of which are visible to the unaided sight of the acutest eye. The little hard-coated plants are often found in fragments, but according to the writer's experience they are broken accidentally, either by being piled on top of each other and so crushed by the cover-glass, or by the rough contact with one another when gathered.

The desmids float freely in the water; many diatoms do the same. Several species of desmids are attached to each other side by side to form long bands; many diatoms are arranged in a similar way. Some desmids are surrounded by a colorless jelly-like coating; so are some diatoms. The desmids never grow on the ends of stems secreted by themselves, and attached to other plants or submerged objects; many diatoms are found growing on the extremities of long colorless and branching stalks like microscopic trees, these stems being fastened to other objects in the water. Some of the commonest diatoms will be found in great abundance growing in this way on the leaves of Myriophyllum. Any object that may apparently be either a desmid or a diatom is not a desmid if it is on the end of a stem of its own formation. Most desmids have the ability to voluntarily change their position. They can move from place to place, as they frequently do when under the microscope, slowly travelling across the field of view in a very interesting way. When mixed with mud or dirt, as they often are when gathered and carried home in a bottle, they will gradually work themselves to the surface and collect in a green film or line on the side of the bottle next the window, whence they can be easily taken by the dipping-tube. Diatoms have a similar power of movement; but they are usually much more active, and their motions more rapid than those of desmids. And while the desmids move stately and slowly in one direction, a diatom may travel quickly half-way across the field of view, and without a moment's hesitation, and without turning round, may at once return by its former path or dart off obliquely. A moving diatom always seems to have important business on hand, and to be anxious to accomplish it. An object, therefore, that may be either a desmid or a diatom is not a desmid if it moves rapidly and changes its course suddenly and quickly. The cause of this motion is in either case a mystery. Many theories have been proposed to explain it, but none are satisfactory. If the reader can discover how the desmids and diatoms move themselves his name will be remembered among naturalists to the end of time.

The surface of a desmid may be smooth, finely striated lengthwise, roughened by minute dots or points, or it may bear several wart-like elevations or spines of different shapes; its edges may be even or notched, prolonged into teeth, or variously cut and divided. It is these ornaments, in connection with the graceful form and the pure homogeneous green color, that make the desmids so attractive to every student of microscopic aquatic life. Fresh-water diatoms occasionally have tooth-like processes, but they are never spine-bearing; yet the markings of their surfaces are among the most exquisite of Nature's handiwork, and the most varied. Dots, hemispherical bosses, hexagons, transverse and longitudinal lines of astonishing fineness, are among their many

4*

surface sculpturings, the delicacy and the closeness of
which defy description. So fine and close together are
the surface lines of some that they are used to test the
good qualities of the best and highest power objectives:
There are no perfectly smooth diatoms, although many
may appear so to a low-power lens; but the splendid
glasses of the best American makers will compel any
diatom to show just how it is marked and roughened.

In each end of many desmids, especially in the crescent-shaped ones, is a small colorless, apparently circular space containing numerous very minute black particles in incessant motion. These little granules, which
are said to be crystals, are sometimes so few that they
can be counted if sufficiently magnified, while in other
individuals they are innumerable. Their motion resembles the swarming of microscopic bees. It can scarcely
be described, but once seen it can never be forgotten.
The spaces containing them are called vacuoles, and are
never present in diatoms. It is true that in some of
the latter, when dying or dead, many minute black particles are visible, dancing and swarming in clusters within the cells, but this is common to many microscopic
creatures after death. In the desmids there is also often
seen a circulation of the protoplasm similar to the cyclósis in the leaf-cells of Anacharis, a movement of the
cell contents never observed, so far as I am aware, in
any diatom. Between the cell-wall and the green coloring matter, the chlorophyl, there seems to be a narrow space filled with colorless protoplasm, and it is here

that the circulation takes place. It is a steady, quite rapid flow, several currents streaming lengthwise up and down the cell, carrying the minute starch grains and other enclosed particles in their course. It has been said that these currents sometimes enter the vacuoles, and that the latter obtain their supply of swarming granules from the particles in the streams; it has also been stated that occasionally one or more of the swarming granules leaves the vacuole, enters the current, and journeys round the cell. These statements are rather doubtful. But with a high-power objective (the one-fifth, for instance) it is not difficult to select a granule, and follow it as the current carries it down one side to the vacuole, where, according to the writer's observation, it never enters, but passes into an ascending current, and continues the round. The vacuoles themselves are visible with a good low-power objective, but to see the swarming granules and the general cyclósis a one-fourth or one-fifth is needed.

In addition to the desmids and diatoms, almost every pond and stream contains other minute plants of interest to the microscopist, called the fresh-water Algæ, which he probably already knows, if not by this name, at least by their general appearance, for they form those green masses floating like a scum on the surface, or soft green clouds attached to sticks and stones and dead leaves. The Algæ often have a disgusting appearance as they collect in thick and heavy patches, but under the microscope they reveal beauty undreamed of. All those slimy, slippery streamers usually so abundant in still water during the summer are Algæ. The beginner need have no trouble to recognize them as Algæ after a little experience, but since he at first may be somewhat uncertain as to which of the three classes of plants his specimen belongs, the following Key has been constructed to aid him. To use it, compare the plant with it in the following way:

Suppose the specimen is a single cell, shaped like a crescent, as described in the first sentence of the Key. The reader will notice (a) at the end, meaning that he shall now seek a description somewhere in the table below with a at the head of the line. Finding three such lines, he reads the first, "Color green," which is the color of the specimen under the microscope; "the plant a floating hollow sphere," which does not describe it, since it is crescent-shaped. He then reads the second "a" line: "Color green, the plant not a hollow sphere," which is of course right, as his plant is not a sphere. The (b) at the end refers to another line below headed by b. There being but one such, the plant must be a desmid; but to learn which of the numerous desmids it is, he turns to Section I. of this chapter, where is another Key to help him find the name of the genus. Again, suppose he obtains a floating mass which, when lifted on the hand or in the dipper, he sees to be a fine, delicate green net. To find the section to which this belongs, read each numbered sentence at the beginning of the Key: 1 will not do, since the specimen is not spherical, crescentic,

nor circular; 2 will not do, because the plant is not in
long threads; 3 and 4 do not describe it, because it is
neither star-shaped nor formed of oval cells with two
bristles on each end; but 5 calls for a green net often
visible to the naked eye, which describes the specimen,
giving the name of its genus, *Hydrodictyon*, and referring the student to the Algæ, Section III. of this chapter. After using this preliminary Key for a few times,
he will be able to decide at a glance through the microscope to which section his specimen belongs.

Key to the Desmids, Diatoms, and Fresh-water Algæ.

1. Plants formed of a single, crescent-shaped, spherical,
   barrel-shaped, oblong and constricted, or circular
   and flattened, cell, sometimes arranged side by side
   in long ribbons, but seldom end to end; color green
   or brown (*a*).
2. Plants formed of many cells arranged end to end in
   long threads; coloring matter usually green, often
   in spiral bands or other patterns on the cell-wall (*d*).
3. Plants formed of several green cells grouped in the
   shape of a flat disk with six to many short blunt star-
   like points; floating free. *Pediástrum* (*Algæ, III.*).
4. Plants formed of two to eight narrowly-oval green
   cells placed side by side, each terminal cell with
   two curved colorless bristles; floating free. *Scenedésmus* (*Algæ, III.*).
5. Plants forming a green net visible to the naked eye.
   Hydrodictyon (*Algæ, III.*).

a. Color green, the plant a floating hollow sphere. *Vólvox* (*Algœ, III.*).

a. Color green, the plant not a hollow sphere (*b*).

a. Color golden-brown (*c*).

b. Cell-wall smooth, rough, warty, or spine-bearing, also soft and flexible; always floating freely, never growing on stems permanently attached to other objects; a vacuole with swarming granules often present in each end. (*Désmids, I.*)

c. Cell-wall marked transversely, often also longitudinally, by lines, smooth bands, or dots; never spine-bearing; cell-wall also hard and brittle; floating freely, or growing on colorless stems permanently attached to other objects. (*Diatoms, II.*)

d. Plants forming cloud-like clusters, long streamers, or scum-like floating masses visible to the naked eye; color bright green or olive, sometimes almost black; the cells under the microscope united end to end to form long, sometimes branching filaments. (*Algœ, III.*)

1. DÉSMIDS.

As the desmids are singly invisible to the naked eye, the student can know what he has gathered only after reaching home, except in those rare instances where the little plants have become congregated together in such quantities that a good pocket-lens will show their forms. I have more than once found *Clostérium* in this profusion, but never any other. The early spring, as early as the last of March or the first of April, in the writer's locality (New Jersey), is the best time of the year to gather them, or indeed any of the Algæ. At that time all these plants seem more vigorous, their vital functions are performed more actively, and the observer is then almost sure to see in some of them the conjugation, or union, of two separate cells and the formation of the spores. This spore formation, however, is much more frequently seen in the thread-like Algæ than in the single-celled desmids.

There are more than four hundred known species of desmids. Perhaps an undue proportion has been included in the subsequent list, but nature offers them so freely and abundantly, and they are so attractive, that they must be their own excuse.

The following Key to the genera is to be used as directed for the "Key to the Desmids, Diatoms, and Fresh-water Algæ," except that when the name of the genus has been found, the reader should then refer to the paragraph on the following pages headed by that name, where he will find one or more species described and figured. Thus, if he has a green half-moon-shaped plant under the microscope, to learn its name turn to this Key, the second line of which describes it, since it is not in ribbons or bands; he then refers to the lines headed by *d*, the first one describing his plant as a "cell more or less crescent-shaped," giving the generic name *Clostérium*, 6 being the number of the paragraph further on in this section of the chapter, where several species are noticed.

Key to Genera of Desmids.

1. In ribbons or narrow bands (*a*).
2. Not in ribbons nor bands (*d*).
     a. In a transparent, jelly-like sheath (*b*).
     a. Not in a jelly-like sheath (*c*).
     b. Each cell with two teeth on each narrow end. *Didymóprium*, 1.
     b. Each cell deeply divided almost into two parts. *Sphærozósma*, 2.
     b. Each cell without teeth and not divided. *Hyalothéca*, 3.
     c. Cells barrel-shaped, the band not twisted. *Bambusína*, 4.
     c. Cells not barrel-shaped, the band twisted. *Desmídium*, 5.
     d. Cell more or less crescent-shaped. *Clostérium*, 6.
     d. Cell cylindrical, spindle-shaped, hour-glass, or dumb-bell shaped (*f*).
     d. Cell flattened, oblong, circular, or often divided into arms (*e*).
     e. Mostly circular or broadly elliptical, often cut and divided by slits and depressions; marginal teeth usually sharp. *Micrastérias*, 7.
     e. Mostly oblong or elliptical, the margin wavy, the depressions rounded. *Euástrum*, 8.

f. Cell constricted in the middle; no arms nor sharp
   spines (*g*).

f. Cell constricted in the middle, with arms or sharp
   spines (*h*).

f. Cell not constricted in the middle; no arms nor
   sharp spines (*i*).

g. Ends notched, cell cylindrical. *Tetmémorus*, 9.

g. Ends not notched, cell cylindrical. *Docídium*,
   10.

g. Ends not notched; cell more or less dumb-bell or
   hour-glass shaped (*l*).

h. Arms three or more, radiating, tipped with one or
   more points. *Staurástrum*, 12.

h. Arms none, spines four, two on each end. *Arthrodésmus*, 14.

h. Arms none; spines several, on the edges; a rounded, truncate, or denticulate tubercle in the centre
   of each semi-cell. *Xanthídium*, 13.

i. Chlorophyl in a spiral band, cell cylindrical. *Spi-
   rotœniä*, 15.

i. Chlorophyl not in a spiral band, cell cylindrical (*k*).

k. Surface roughened by tooth-like elevations. *Triplocéras*, 16.

k. Surface smooth, ends rounded, neither divided nor
   notched. *Pénium*, 17.

l. End view three to six or more angular (*m*).

l. End view not angular, margins smooth, dentate,
   or crenate, without spines; ends always entire.
   Cosmárium, 11.

m. Angles obtuse, acute, or with horn-like prolongations. *Staurástrum,* 12.*

1. Didymóprium.

Each cell in the band longer than broad; two rounded or angular teeth on each narrow end; case or sheath distinct, colorless. *D. Grevíllii,* Fig. 17.

Fig. 17.—Didymóprium Grevíllii.

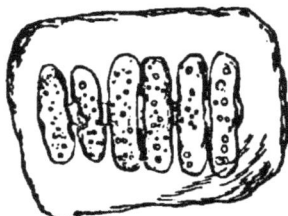

Fig. 18.—Sphærozósma pulchra.

2. Sphærozósma.

Each cell in the band about twice as long as broad, divided on both ends almost to the middle; sheath large, colorless. Three cells are shown in the figure. *S. púlchra,* Fig. 18.

3. Hyalothéca.

The ribbons are often very long, and the narrow ends of each cell are sometimes slightly constricted, as shown in the lower part of the figure, but the depression is never deep enough to form teeth; sheath colorless. *H. dissíliens,* Fig. 19.

Fig. 19.—Hyalothéca dissíliens.

Fig. 20.—Bambusina Brebissónii.

4. Bambusína.

The cells somewhat resemble barrels or casks joined

end to end, with two narrow hoop-like elevations around the middle of each. *B. Brebissónii*, Fig. 20.

5. DESMÍDIUM.

The twisted appearance of the band is due to the fact that each cell is triangular, as may sometimes be seen when they break apart and float over on end, but the three angles are not all in the same line, each cell being turned a little to one side. When the side of the band is looked at, it is these angles that are seen like a dark oblique or zigzag line traversing the ribbon. Each cell is slightly toothed on both the narrow ends. Common. *D. Swártzii*, Fig. 21.

Fig. 21. — Desmidium Swartzii.

6. CLOSTÉRIUM (Figs. 22 to 31).

All the species of this genus are more or less crescent-shaped, some being more curved than others, but none having exactly straight sides. In each end of almost every one will be seen a clear circular vacuole containing many small, dark, swarming granules. These have already been referred to, as has the movement of the protoplasm between the cell-wall and the layer of green coloring matter. *Clostérium* is the only desmid in which this cyclosis can be seen easily, if it ever occurs in others. There are thirty-five species of the genus, the following being some of the commonest. The most convex margin is called the " back ;" the concave border the " ventrum."

Some Species of Clostérium.

1. Ends not lengthened out into a colorless beak (*a*).
2. Ends lengthened out into a colorless beak (*f*).
   - *a.* Back slightly convex, the whole cell slightly curved (*b*).
   - *a.* Back strongly convex, ventrum nearly straight (*c*).
   - *a.* Back strongly convex, ventrum strongly concave, with a central enlargement (*d*).
   - *a.* Back strongly convex, ventrum without a central enlargement (*e*).
   - *b.* Ventrum nearly straight; vacuoles close to the rounded ends; fifteen or twenty chlorophyl globules in a central longitudinal row in each semi-cell. *C. lineátum,* Fig. 22.

Fig. 22.—Clostérium lineátum.

   - *b.* Ventrum nearly straight; body tapering towards the rounded, sometimes curved, ends; vacuoles small, often scarcely visible. *C. juncídum,* Fig. 23.

Fig. 23.—Clostérium juncidum.

   - *b.* Ventrum and back equally curved; ends tapering; ten to fourteen chlorophyl globules in a central, longitudinal row in each semi-cell; vacuoles very small. *C. acerósum,* Fig. 24.
   - *c.* Ends rounded; chlorophyl often arranged in nar-

Fig. 24.—Clostérium acerósum.

row, longitudinal bands; chlorophyl globules numerous; vacuoles near the ends; cell very large.
C. Lúnula, Fig. 25.

Fig. 25.—Clostérium Lúnula.

d. Ends rounded; chlorophyl often arranged in narrow, longitudinal bands; chlorophyl globules often numerous; vacuoles close to the ends.   *C.
Ehrenbergii*, Fig. 26.

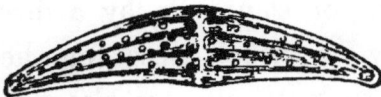

Fig. 26.—Clostérium Ehreubergii.   Fig. 27.—Clostérium acumínátum.

e. Large, crescent-shaped; centre broad, ends acute,
vacuoles small.   *C. acuminátum*, Fig. 27.

e. Small, crescent-shaped, distance between the ends
about ten times the central diameter; centre narrow, vacuoles indistinct.   *C. Diánæ*, Fig. 28.

Fig. 28.—Clostérium Diánæ.   Fig. 29.—Clostérium Véuus.

e. Very small, crescent-shaped, eight to twelve times
as long as broad; centre narrow, ends sharp,
vacuoles distinct.   *C. Vénus*, Fig. 29.

f. Each beak about as long as the green body, sometimes shorter; whole cell slightly curved, vacuoles usually indistinct.   *C. rostrátum,* Fig. 30.

Fig. 30.—Clostérium rostrátum.

f. Each beak extremely fine, longer than the spindle-
shaped green body, the tips alone curved.   *C. setáceum,* Fig. 31.

Fig. 31.—Clostérium setáceum.

7. MICRASTÉRIAS (Figs. 32 to 39).

Each Micrastérias is divided across the middle into
two equal and similar halves, or semi-cells, by a deep
slit, the sides of which may be either close together
or somewhat separated. Both semi-cells are also very
much slit and notched, but both in the same way, the
description of one half, therefore, applying equally well
to the other. There are forty-two species of the genus.
The beginner must expect to find many forms not included in this list, which contains only some of the most
common in the writer's vicinity.

Some Species of Micrastérias.

1. More or less circular in outline (*a*).
2. Not circular; divided into radiating arms (*b*).
3. Not circular; not divided into arms; central slit
gaping (*c*).

a. Each semi-cell divided by four deep cuts into one
end and four side
lobes, each side-lobe
divided by a short-
er cut into two sec-
tions, each section
by a still shorter
cut divided into
two divisions, each
division by a yet
shorter cut divided
into two parts, and
each part with two

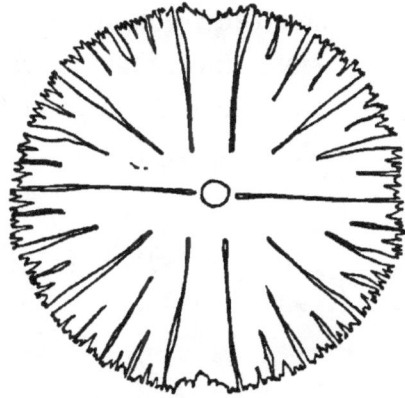

Fig. 32.—Micrastérias radiósa.

teeth.   Desmid very large.   *M. radiósa*, Fig. 32.

a. Each semi-cell divided by four cuts into one end
and four side-lobes, each side-lobe divided by a
shorter cut into two parts, and each part with
two teeth.   *M. rotáta*, Fig. 33.

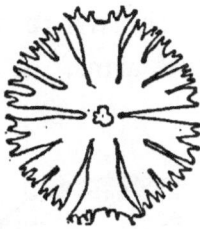

Fig. 33.—Micrastérias rotáta.

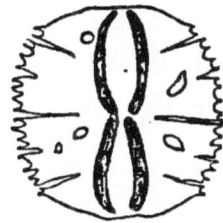

Fig. 34.—Micrastérias truncáta.

a. Each semi-cell divided by two cuts into one end
and two side-lobes, each side-lobe by a shorter cut
divided into two parts, and each part with two

teeth.   End-lobes broad, often with two teeth on
each end.   *M. truncáta*, Fig. 34.

b. Each semi-cell divided by deep rounded depressions into four tapering, slightly curved arms,
the whole desmid having eight undivided arms.
M. arcuáta, Fig. 35.

Fig. 35.—Micras-   Fig. 36.—Micrastérias   Fig. 37.—Micrastérias Kit-
térias arcuáta.   dichótoma.   chélii.

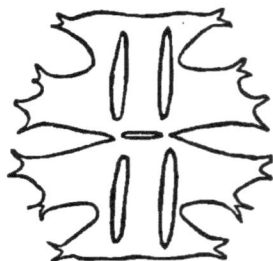

b. Each semi-cell divided by two acute depressions
into one end and two side lobes, each side-lobe
divided by an acute depression into two short
parts, each part divided by an acute depression
into two short arms, and each arm with two teeth;
arms of the end-lobes each with two teeth; the
whole desmid with twenty short arms. *M. dichótoma*, Fig. 36.

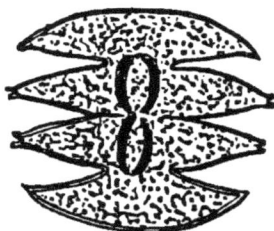

Fig. 38.—Micrastérias óscilans.   Fig. 39.—Micrastérias láticeps.

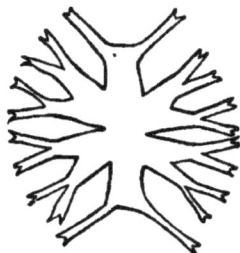

c. Divided into one end and two side lobes (*d*).

d. Side-lobes divided by a shallow notch into two parts extending beyond the end-lobes, each part with two teeth on both ends. *M. Kitchélii*, Fig. 37.

d. Side-lobes not divided into two parts, but extending beyond the end-lobes. *M. óscitans*, Fig. 38.

d. Side-lobes not divided into two parts, not extending beyond the end-lobes. *M. láticeps*, Fig. 39.

8. EUÁSTRUM (Figs. 40, 41, 42).

Euástrum is divided into two halves by a central slit across the middle, but the cell is never circular as in Micrasterias, and the margins are wavy, never sharply toothed. The ends are usually notched. There are about forty species.

1. Each half oblong; an end-lobe present in both halves, and formed by a short rounded cut on each side. *E. crássum*, Fig. 40.

Fig. 40.—Euástrum crás-   Fig. 41.—Euástrum   Fig. 42.—Euástrum
sum.   didélta.   ansátum.

2. Each half somewhat triangular, without distinct end-lobes (*a*).

5

a. Sides wavy, gradually expanding towards the central cut. *E. didélta,* Fig. 41.

a. Sides hardly wavy, suddenly expanding towards the central cut. Small. *E. ansátum,* Fig. 42.

9. TETMÉMORUS (Figs. 43, 44).

1. Widest in the middle, the ends tapering. *T. granulátus,* Fig. 43.

Fig. 43.—Tetmémorus granulátus.   Fig. 44.—Tetmémorus Brebissónil.

2. Not widest in the middle, the ends not tapering. *T. Brebissónii,* Fig. 44.

10. DOCÍDIUM (Figs. 45, 46).

1. A rounded enlargement on each side of the central constriction. *D. Báculum,* Fig. 45.

Fig. 45.—Docidium Báculum.   Fig. 46.—Docidium crenulátum.

2. Two or more small enlargements on each side of the central constriction, giving the edges a wavy appearance. *D. crenulátum,* Fig. 46.

11. COSMÁRIUM (Figs. 47, 48, 49, 50).

The ends of Cosmárium are never notched nor incised. They may be, and often are, rough or warty, but the ends are always entire. There are about one hundred species. The following are common:

1. Surface smooth; cell less than twice as long as broad,

the two semi-cells evenly rounded. *C. Rálfsii*, Fig. 47.

2. Surface smooth; cell about twice as long as broad, the margins of each semi-cell slightly sloping towards the flattened ends. *C. pyramidátum*, Fig. 48.

Fig. 47.—Cosmárium Rálfsii.

Fig. 48.—Cosmárium pyramidátum.

Fig. 49.—Cosmárium margaritiferum.

Fig. 50.—Cosmárium Brebissónii.

3. Surface roughened by rounded, pearly elevations. *C. margaritiférum*, Fig. 49.

4. Surface roughened by minute sharp points. *C. Brebissónii*, Fig. 50.

12. STAURÁSTRUM (Figs. 51, 52, 53, 54).

In front view, or in the position in which the desmids usually lie naturally, Staurastrum resembles Cosmarium, but in end view it is always angular. It is sometimes rather troublesome to get Staurastrum, or indeed any other desmid, tilted up on end so that it can be examined in that situation, but in a moderately deep cell, with considerable water and a low-power objective, it can usually be turned over by gently tapping and pressing the cover-glass with a needle.

Staurastrum is a large genus, containing about one hundred and twenty species.

1. Cell dumb-bell shaped; without arms; surface rough-

ened by small elevations. *St. punctulátum*, Fig. 51.

2. Cell not dumb-bell shaped; with arms (*a*).

  a. Cell triangular in end view, the angles toothed; arms in a cluster of about three on the end of the cell, their ends toothed. *St. furcigerum*, Fig. 52.

Fig. 51.—Stanrástrum punctulátum.   Fig. 52.—Stanrástrum furcigerum.   Fig. 53.—Stanrástrum grácile.   Fig. 54.—Stanrástrum macrócerum.

  a. Cell triangular in end view, the angles prolonged as narrow arms, the ends of which are three-toothed; surface roughened. *St. grácile*, Fig. 53.

  a. Cell with six or seven radiating arms, their ends three-toothed. *St. macrócerum*, Fig. 54.

13. XANTHÍDIUM (Figs. 55, 56).

The cells bear near both ends a prominence or tubercle that may be rounded and smooth, truncate, or apparently encircled by small beads.

1. Cell about twice as long as broad; spines short, their ends irregularly toothed; tubercles circular, beaded. This is the only species with toothed spines. *X. armátum*, Fig. 55.

2. Cell not twice as long as wide, each half somewhat kidney-shaped; spines in four or six pairs on each

semi-cell, not divided nor toothed, but often curved; tubercle a curved row of bead-like elevations. *X. antilopæum*, Fig. 56.

Fig. 55.—Xanthidium armátum.   Fig. 56.—Xanthidium antilopæum.

14. ARTHRODÉSMUS (Figs. 57, 58).

1. Spines on the same side curving or spreading from each other; surface smooth. *A. incus*, Fig. 57.
2. Spines on the same side curving towards each other; surface smooth. *A. convérgens*, Fig. 58.

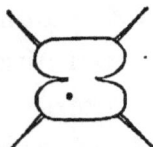  

Fig. 57.—Arthrodésmus   Fig. 58.—Arthrodésmus   Fig. 59.—Spirotænia con-
incus.   convérgens.   densáta.

15. SPIROTÆNIA.

Ends rounded; spiral band closely wound. *S. condensúta*, Fig. 59.

16. TRIPLOCÉRAS.

Surface roughened by small projections arranged in rows around the cell, their tips notched or finely toothed; cell twelve to twenty times as long as broad. *T. verticillátum*, Fig. 60.

Fig. 60.—Triplocéras verticillátum.

17. Pénium.

Cylindrical; ends rounded, surface smooth. *P. Brebissónii*, Fig. 61.

If it is desired to preserve any of the desmids or Algæ, the following solution will be found to be an ex-

Fig. 61.—Pénium Breblssónll.

cellent medium. In it the plants retain their green color, and the cell contents have less tendency to shrink from the cell-wall than with any other of the many media often recommended. Any druggist can make the solution. It is composed as follows: Camphorated water and distilled water, of each 50 grammes; glacial acetic acid, 0.5 gramme; crystallized chloride of copper and crystallized nitrate of copper, of each 2 grammes; dissolve and filter.

The plants should be placed in a cell made of shellac, a few drops of this preservative copper-solution added, and the cover fastened down with shellac. If any other cement, except perhaps Brown's rubber cement, is used with this solution it will inevitably run under and ruin the preparation.

If the beginner should find the desmids so attractive that he desires to study them rather than to learn the names and appearance of a few of the commonest, he should refer to the Rev. Francis Wolle's excellent book entitled "The Desmids of the United States."

II. DIATOMS.

For a long time there was much discussion as to the animal or vegetable nature of the diatoms, but that they are plants is now the general belief. Their peculiar motion was one great reason for classing them among the animals, although some undoubted plants have even a more rapid movement.

No class of microscopic objects, except, perhaps, the Infusoria, is so abundant. No ditch or pond is without them. No pool is too small to harbor them; even a depression made by a cow's hoof in a wet meadow soon becomes a home for them. They will probably form some of the first things to attract the attention of the beginner in the use of the microscope.

Their shape is as varied as their number is great, and their hard and glass-like surface is most beautifully lined and dotted, and sculptured in delicate tracery. Most plants are comparatively soft, but the diatoms are noteworthy for the hard case enclosing the semi-fluid, yellowish-brown contents, a case that is indestructible. It may be heated to redness, it may be boiled in strong acids and alkalies, and at the end be as it was before, as gracefully formed and as beautifully marked. Indeed, to properly study the diatoms they should be treated by some method to destroy the coloring matter often obscuring the surface markings for which they are chiefly valued. For the beginner, however, who desires only to recognize a diatom when he meets with one in the field of his microscope, and to learn its name, if possible, such preparation is unnecessary.

Diatoms are also peculiar in their structure. In this they have often been compared to a pill-box. The diatom is formed of two parts called valves, one of which may be likened to the pill-box proper, and the other to the lid, since it slips over the edge of the lower valve. The entire box-like diatom is called the frustule; the surfaces of the upper and lower valves are usually marked and shaped alike, and are called the sides. But when the frustule happens to be turned so that the narrowest part, or that part corresponding to the thickness of the pill-box, and called the front, is towards the observer, then the shape is so different from that of the valves as to puzzle the beginner. If in doubt about the position, gently tap the cover-glass with a needle, when the frustule will generally roll over on its broad side. This seems somewhat bewildering at first, but there is no difficulty if it is borne in mind that the thickness of the pill-box corresponds to the *front* of the frustule, and the broad surfaces of the lid and bottom to the sides of the *valves*.

In addition to the ordinary markings on the valves—that is, the transverse lines which are sometimes so coarse that they are called ribs—each valve frequently bears a line or narrow smooth band down the middle, while at each end and at the centre there is often a small rounded spot resembling a circular space, but being in reality an elevation, called a nodule.

In remote ages diatoms existed in even greater numbers than at present. Immense beds of fossil frustules are found in many parts of the world, especially in our own country. In Maryland and in New Jersey diatomaceous earth is obtained containing exquisite forms. In Virginia a certain deposit is especially renowned, since it is eighteen feet thick and underlies the city of Richmond. This has afforded the student some of the rarest and most valued frustules, or valves, for the frustule, before it can be properly studied, must be separated into its two valves. To have produced such a mass they must have existed in incalculable numbers in a great body of water where, dying, and sinking to the bottom year after year, their skeletons accumulated as others continued to fall. To appreciate the probable length of time, as well as the number of diatoms, needed to make such a deposit, it is only necessary to know that a single frustule is seldom thicker than the one ten-thousandth of an inch.

At the present day living diatoms are often found in large numbers forming a yellowish-brown film on the mud in shallow water. In such cases it is no trouble to skim them up and so gather them. Usually, however, the beginner will first see them floating freely about his slide, or attached to various plants. But few are visible to the naked eye except when collected in great masses, and only then as brownish patches; the individual valves are seldom seen without the microscope, and then only to the most acute and best educated eye.

5*

They are difficult to study and to name. To properly examine them demands the highest power objectives of the best construction, and a skill in the use of the microscope and accessary optical apparatus not at the beginner's command. Much has been written about them, but the literature of the subject is so widely scattered through the scientific magazines that only those who make a special study of the diatoms can hope to have it in their libraries. But the beginner need not despair. With ease he can learn to recognize a diatom whenever seen, and to know the names of the commonest forms, and this is all he will care to learn at first. Yet he will find it a satisfaction to be able to say to a friend, "That is a diatom," and to explain its box-like structure.

The following Key has been made to assist the beginner in ascertaining the names of a few of the commonest fresh-water forms. It is impossible to include even a tithe of the plants, and the beginner will surely find many not mentioned in the succeeding list, but from the brownish color, the movements common to so many, and the hard, dotted, lined, and sculptured valves, he can readily know them as members of the *Diatomaceæ* after he has seen and recognized one. More than this he can scarcely hope to do.

The brown coloring matter will often be seen contracted to a narrow strip or to a spot at each end, and very frequently the frustule will be entirely colorless. Diatoms are the favorite food of many microscopic animals, which absorb the cell contents, often leaving the hard and indigestible valves colorless, but otherwise unchanged.

Key to Genera of Diatoms.

1. Growing in bands or ribbons (*a*).
2. Growing on colorless stems or in a jelly-tube (*c*).
3. Growing with their concave sides attached to other plants (*e*).
4. Free-swimming (*f*).

    a. Band curved or coiled. *Merídion*, 1.

    a. Band zigzag; frustules attached together by the corners. *Diátoma*, 2.

    a. Band uneven, frustules long, narrow, rapidly sliding on each other. *Bacillária*, 3.

    a. Band straight, or nearly so, edges even, frustules motionless (*b*).

    b. Each frustule six times as long as broad. *Fragelária*, 4.

    b. Each frustule twice as long as broad. *Himantidium*, 5.

    c. In a narrow jelly-tube; valves boat-shaped. *Encyonéma*, 6.

    c. On the ends of colorless stems (*d*).

    d. Valves boat-shaped. *Cocconéma*, 7.

    d. Valves wedge-shaped. *Gomphonéma*, 8.

    e. Valve six to seven times as long as broad. *Epithémia*, 9.

    e. Valve oval, nearly as long as broad. *Cocconéis*, 10.

f. Valve not curved nor S-shaped (*g*).

f. Valve in side view arched, the convex edges scalloped. *Eunótia*, 11.

f. Valve long S-shaped. *Pleurosigma*, 12.

f. Valve boat - shaped, the ribs conspicuous. *Surirélla*, 13.

f. Valve boat-shaped, ribs none. *Navícula*, 14.

g. Valve strongly ribbed ; a nodule at each end and
in the centre. *Pinnulária*, 15.

g. Valve not ribbed ; with a central longitudinal, and
. a transverse smooth band. *Stauronéis*, 16.

1. MERÍDION CIRCULÁRE (Fig. 62).

Valves wedge - shaped, transverse lines indistinct,
bands spiral, often broken into small curved sections
(Fig. 62).

Fig. 62.—Merídion circuláre.   Fig. 63.—Diátoma vulgáre.

2. DIÁTOMA VULGÁRE (Fig. 63).

Frustules oblong, the four angles right-angles, band
often attached to aquatic plants, easily separable into
its component frustules (Fig. 63).

3. BACILLÁRIA (Fig. 64).

Frustules long and narrow, united laterally, freely

and rapidly sliding backward and forward over each other; free-swimming (Fig. 64).

This is probably one of the most interesting of the common fresh-water diatoms, on account of its strange movements. When quiet, as it probably will be immediately after being placed on the slide, the band will somewhat resemble a row of fence pickets lying closely side by side. Suddenly each picket shoots forward until they are all nearly end to end, the band becoming a long irregular line, and quite as suddenly closing together again. This alternate backward and forward gliding is continued until the diatoms become apparently exhausted, or the oxygen in the water is consumed. What prevents one frustule from slipping off the end of the other is not known; indeed the cause of the entire performance can only be guessed at. All the species of the genus Bacillaria are said to live in salt water. The form which I have ventured to identify as a sweet-water variety of *B. paradóxa* is not uncommon in fresh-water ponds.

Fig. 64.—Bacilláris.

4. Fragelária capucina (Figs. 65 and 65*a*).

Frustules very narrow, never wedge-shaped, band long. Fig. 65 shows the band of united frustules; Fig. 65*a* the appearance of a single valve more highly magnified. The ends of the valves are somewhat wedge-shaped.

Figs. 65 and 65*a*.—Fragelária capucina.

5. Himantídium pectinále (Fig. 66).

Frustules much wider than the preceding, transverse lines distinct on both sides of a narrow central smooth space (Fig. 66).

Fig. 66.—Ilimantidium pectinále.

Fig. 67.—Encyonéma paradóxa.

6. Encyonéma paradóxa (Fig. 67).

The jelly-tubes are usually very slightly, if at all, branched, the frustules commonly arranged in longitudinal lines within the tubes, which are attached to other plants (Fig. 67).

7. Cocconéma lanceoláta (Figs. 68 and 68a).

Stems often much branched, attached to aquatic plants and other submerged objects, frustules on the ends of the branches, in side view (valves) slightly curved with a median longitudinal line, a nodule at each end and in the centre (Fig. 68). The frustules are often found separated from the stems and floating freely, when they are usually seen in side view. Fig. 68a shows a single valve highly magnified.

Fig. 68 and 68a.
—Cocconéma
lanceoláta.

8. Gomphonéma acumináta (Fig. 69).

Stems often much branched, but frequently found unbranched; attached to other plants; frustules slightly swollen in

the centre, the narrowest end of the wedge
attached to the stem (Fig. 69).

9. EPITHÉMIA TÚRGIDA (Fig. 70).

Valves curved or bent, transverse lines
coarse and conspicuous (Fig. 70).

10. COCCONÉIS PEDÍCULUS (Fig. 71).

Valves oval, with a line down the centre and a small nodule in the middle; at-

Fig. 69.—Gompho-
néma acumináta.

tached by one valve to aquatic plants, especially to the
leaves of *Anácharis* (Fig. 71).

Fig. 70.—Epithémia tur-
gida.

Fig. 71.—Cocconéis
pediculus.

Fig. 72.—Eunótia
tetráodon.

11. EUNÓTIA TETRÁODON (Fig. 72).

Valves curved, a small nodule at each end of the
concave margin; the convex border apparently scalloped, but in reality bearing four or more rounded
ridges (Fig. 72).

12. PLEUROSÍGMA (Fig. 73).

Valves long S-shaped, widest in the middle and tapering to both ends, one of which curves towards the

right-hand side, the other towards the left-hand (Fig. 73). A narrow S-shaped line extends down the centre of the valve, with a nodule in the middle. There are a large number of species of this genus, all of which may be known by this peculiar and beautiful curvature of the sides, the word Pleurosígma meaning S-shaped sides. The valves are very finely striated, the lines being remarkably close together, and demanding a comparatively high-power objective of excellent construction to see them. The valves are therefore often used to test the good qualities of certain objectives. To the beginner, however, all the Pleurosigmas will probably appear to be smooth. The species most frequently used as a test is *P. angulátum*, a salt-water form.

Fig. 73.—Pleurosigma.

13. SURIRÉLLA SPLÉNDIDA (Fig. 74).

The ribs are large and conspicuous, the spaces between them seeming to be lower than the edges of the valves, thus giving the latter the appearance of having a narrow wing around the margin (Fig. 74). The members of this genus are also used as test-objects, the one most commonly employed being a marine species.

Fig. 74.—Surirélla spléndida.

14. NAVÍCULA CUSPIDÁTA (Fig. 75).

Valves widest in the centre, tapering with straight

margins to each end; a straight line down
the middle with a central nodule (Fig. 75).

Fig. 75.—Navicula
cuspidáta.

15. PINNULÁRIA (Figs. 76 and 77).

1. Sides of the valves parallel, the ends
and centre somewhat swollen; trans-
verse ribs large and conspicuous; a line down the
middle, with a nodule at each end and at the centre; frustule large. Common. *P. major*, Fig. 76.

Fig. 76.—Pinnulária major.

Fig. 77.—Pinnulária viridis.

2. Sides of the valves slightly convex, the ends and the
centre not swollen; ribs large and conspicuous; a
line down the middle, with a nodule at each end
and one in the centre; frustule smaller than the
preceding. It is named green (*viri-
dis*), probably because it is always
brown. *P. víridis*, Fig. 77.

Fig. 78.—Stauro-
néis phœnocén-
teron.

16. STAURONÉIS PHŒNOCÉNTERON (Fig. 78).

Valves widest in the middle, tapering
with curved sides to both ends; central and transverse
bands smooth and conspicuous. Common (Fig. 78).

III. FRESH-WATER ALGÆ.

The Algæ often collect together and form green
clouds in the water or a scum-like growth on the sur-

face. Frequently, however, the student will find isolated filaments under his microscope, and not know how
they were placed there, or he will find single threads
adherent to other objects which he may be examining.
The color of the visible masses is usually bright green ;
it may be brownish if the plants are in fruit, or the natural tint of the individual alga may be brownish or
purplish. Many species are coated with a mucous or
slimy material that makes them very slippery and difficult to handle, or to remove from the water unless a
dipper or spoon be used.

They are seldom found in any abundance in deep
water. They seem to prefer shallow ponds and slowly
flowing streams, where they may have plenty of warmth
and light. Few of the species are free-swimming.
Most kinds are adherent to leaves, stones, or sticks in
the water; some form feathery clusters of branching filaments, others surround themselves by little balls of
translucent jelly often attached to leaves of grass or to
other submerged objects.

The following have been partially described in the
Key on page 71.

1. SCENEDÉSMUS (Fig. 79).

The cells are usually four, attached together by their
long sides. The spines on the narrow ends of the two
terminal cells are curved towards each other, and a spine
sometimes grows from the centre of one of the middle
cells. The plant is quite common. *S. quadricáuda*,
Fig. 79.

2. Pediástrum (Fig. 80).

The green cells are usually so arranged as to leave narrow colorless bands between them, and occasionally, in those species formed of a great number of adherent

Fig. 79.—Scenedésmus quadricáuda.

Fig. 80.—Pediástrum granulátum.

cells, several apparently empty colorless spaces are scattered about the disk. In the latter cases the colorless marginal teeth are often very numerous, but they are usually more or less conspicuously arranged in twos. In the species here figured the marginal teeth are generally twelve in number. *P. granulátum,* Fig. 80.

3. Vólvox.

A small sphere continually in movement, rolling through the water in a very graceful manner, its surface studded with green points. Under a low power it seems like a hollow globe, and the cause of the motion is a mystery; but the ¼ inch objective, when the Volvox moves slowly or rests, shows that each green point bears two fine cilia or little hairs continually vibrating and lashing the water. It is from their vibrations that the Vólvox receives its rolling motion. The deep green balls often seen within the globe are young plants in different stages of development. When matured the mother-globe is broken and the young plants float out and roll away through the water, revolving as they are often seen to do even before leaving the parent.

The water in some localities is, in June, so filled with these rolling globes that it is colored green by them, and when the collecting-bottle is held against the light they become visible to a sharp eye as small pale-green spheres. The diameter of a full-grown plant is about one-fiftieth of an inch. *V. globátor.*

4. HYDRODÍCTYON (Fig. 81).

A yellowish-green scum is sometimes seen on the water, which, when spread out over the fingers, proves to

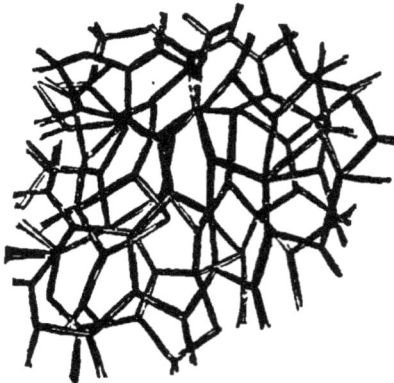

Fig. 81.—Hydrodictyon utriculátum.

be a net of delicate green meshes. ·It may grow to ten or twelve inches in length, and form floating masses several inches in thickness. The nets are composed of narrow short cylindrical cells. Under a low power they are remarkably beautiful. The figure shows a part of a net. *H. utriculátum,* Fig. 81.

The masses which the Algæ form are usually composed of great numbers of long threads, commonly called filaments, and matted together, probably by their rapid growth, among other causes. Each filament is built up of many cells attached to each other by their narrow ends, the single filament being considered a single and entire plant.   They have no roots, although they may fasten themselves by one end to submerged objects.. Some are simple, straight, or curved cellular threads; some give off branches which generally resemble the main plant or stem.   Their color is usually some shade of green, although a few purplish and brownish ones are known.   The following is a key to those genera referred to in this book.

Key to Genera of Fresh-water Algæ.

1. Color brownish-green, bluish, or olive (*a*).
2. Color pure green (*d*).
   a. Filaments branched (*b*).
   a. Filaments not branched (*c*).
   b. Branches with many, whorled, moniliform threads; plant slippery. *Batrachospérmum*, 1.
   c. Moniliform, with larger scattered spherical cells. *Anabœna*, 2.
   c. Not moniliform; bluish green; twisting, bending, gliding. *Oscillária*, 3.
   d. Green color in one or more spiral bands in each cell. *Spirogýra*, 4.
   d. Green color in two star-shaped masses in each cell. *Zygnéma*, 5.
   d. Green color not in patterns (*e*).
   e. Terminal cells with a colorless, hair-like bristle (*f*).
   e. Terminal cells without a bristle. *Vauchéria*, 6.

f. Forming small green, visible, jelly - like masses. *Chœtóphora,* 7.

f. Not forming jelly-like masses (*g*).

g. Cells of the branches green, those of the stem larger, with a transverse green band. *Draparnáldia,* 8.

g. Cells of branches and stem green; bristles with a swollen base. *Bulbochœte,* 9.

1. BATRACHOSPÉRMUM (Fig. 82).

The plant often grows abundantly, attached to submerged objects, in springs, ditches, and ponds. It varies in length from an inch or less to one or two feet, and in color it may be bluish-green, brownish, or purplish. It is covered with a mucous substance that makes it very slippery and difficult to handle. It is much branched, and the branches bear many short threads in whorls, each thread plentifully beaded. The whorls are often so close together that the entire plant as it floats beneath the water seems to be a string of little balls. Under the microscope each moniliform thread is often seen to be terminated by a colorless hair-like bristle. This, however, is not always present. With a high power the ends of the beaded filaments seem to run down the main stem in long, narrow, almost colorless cells. *B. monilifórme,* Fig. 82.

Fig. 82.—Batrachospérmum monilifórme.

2. ANABÆNA (Fig. 83).

Filaments moniliform, freely floating, the cells spherical, the larger scattered ones globular, yellowish.   The

Fig. 83.—Anabæna.

filaments are often curved, and surrounded by a delicate mucous material.   There are several species which closely resemble each other (Fig. 83).

3. OSCILLÁRIA (Fig. 84).

These plants are found almost everywhere in the water.   They often form thick floating mats of a dark purplish almost blackish color, or they are entangled among other plants in a dark green film.   Under the microscope they consist of filaments composed of very many short cells that vary a good deal in width according to the species, of which there are several.   They can usually be known by the bluish-green color and their characteristic motions.   Some are like straight rods of cells bending slowly from side to side; others twist and writhe and coil themselves into circles, only to slowly uncoil and repeat the movements.   Some glide slowly forward, the tip end gradually bending and curving.   The move-

Fig. 84.—Oscillária.

ments, when the plants are in a healthy condition, are incessant.   The beginner need never be at a loss to recognize one of the several species of *Oscillária*.   Three forms are shown in the figure.   (Fig. 84.)

4. SPIROGÝRA (Figs. 85, 86).

The *Spirogyræ* are easily recognized by the beautiful spiral bands of green within each cell, as shown in Fig. 85. There may be one, two, or several of these spirals

Fig. 85.—Spirogýra.

winding around the cell-wall, the number helping to determine the species, of which there are many. The plants usually grow in masses, and especially form those soft green clouds apparently floating in the water. They are often attached to submerged objects, but almost as often free.

Their manner of producing spores is remarkable, but not confined to them, as other Algæ have a similar method. The cells of two filaments lying side by side begin, usually at the same time, to throw out from those sides nearest each other a narrow tube. These tubes meet and grow together, so that the two filaments soon resemble a ladder, the original filaments forming the sides, the tubes being the rounds. The coloring matter then falls away from the cell-walls, and the entire contents of the cells of one filament pass through the rungs of this living ladder into the opposite cells, where the contents of both mingle. From this mixture the spore is formed,

Fig. 86.—Spirogýra in conjugation; with spores.

one in each cell, and is, when ripe, oval and dark brown. This conjugation, as it is called, and the resulting spores are shown in Fig. 86.   The plants are often found in this condition in June and July.

5. ZYGNÉMA (Fig. 87).

This usually floats unattached.   The cells are rather wide and short, the internal stellate masses being a deep green in color.   The formation of the spores resembles that of Spirogýra.

Fig. 87.—Zygnéma insigne..

It is found in conjugation in April.   *Z. insigne*, Fig. 87.

6. VAUCHÉRIA (Fig. 88).

A deep green mat growing on the mud in shallow water, and resembling felt both to touch and sight, will usually prove to be *Vauchéria*.   The filaments are very long, with few branches.   The green matter is diffused over the cell-wall, and when the latter is broken, flows out and often forms green globules.   The spores are produced in two ways, both of which the beginner will see, as they are not rare early in the season.   In one the end of a filament enlarges and becomes club-shaped, while a partition grows across it near the handle of the club.   The contents of this new cell become very dark, opaque, and hardened.   The free end of the cell then breaks, and the spore slowly passes out, being squeezed into an hour-glass shape as it does so.   No sooner is it free than it is off like a flash, being covered by cilia.   But it soon settles down, and finally develops into a fil-

6

ament like the parent. In the other method the filament produces from the side, as shown in Fig. 88, a

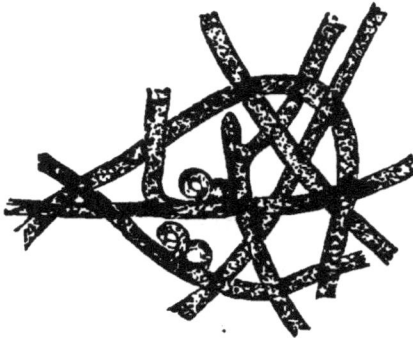

Fig. 88.—Vaucheria.

small oval cell, and near
it a narrow curved or
coiled tube.　Presently
the free ends of each of
these cells open, and the
contents of the tube pass
into the oval cell, in
which a spore without
cilia is finally formed.
This spore is said to fall
in the mud and to remain unchanged for many months,
sometimes all winter, but at last developing into another
Vaucheria. In some of the species the oval cells are
several in a cluster, and the whole, with the coiled tube,
is raised above the filament on the end of a short stem.

7. Chætóphora (Fig. 89).

The light green jelly-like masses into which this Alga
grows are found attached to submerged leaves of grass,
twigs, or other small objects. They are often almost
spherical, varying in size from that of a pin-head to that
of a marble. The surface is smooth, and so slippery that
to pick up one of these *Chætóphora* jellies is next to impossible. The plant within the jelly is formed of fine
branching filaments usually radiating in all directions
from a common centre, the branches being shorter and
most numerous near the surface of the gelatinous mass,

their ends bearing a fine, colorless hair or bristle. Under a low-power objective the plant, if carefully flat-

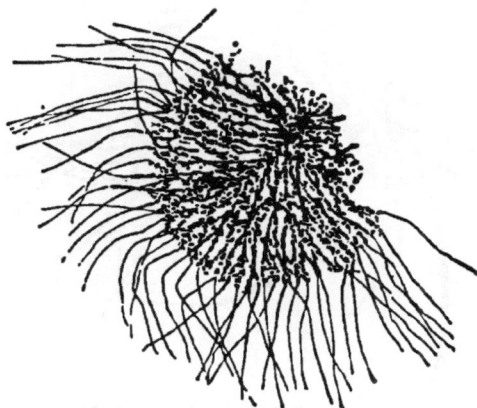

Fig. 89.—Chætóphora élegans.

tened out, is very beautiful. It is justly named "elegant." *Ch. élegans*, Fig. 89.

8. DRAPARNÁLDIA (Fig. 90).

There need be no trouble in recognizing this Alga. It grows attached to many objects, the fine branches giving it a delicate feathery appearance to the naked eye. Under the microscope it is seen to be much branched, the branches being arranged in clusters, and formed of cells smaller than those of the main stem, and filled with chlorophyl, while each terminal cell is ended by a long, colorless hair. The cells of the stem are but little longer than wide, and are colorless except for a narrow, light green chlorophyl band surrounding the centre. *D. glomeráta*, Fig. 90.

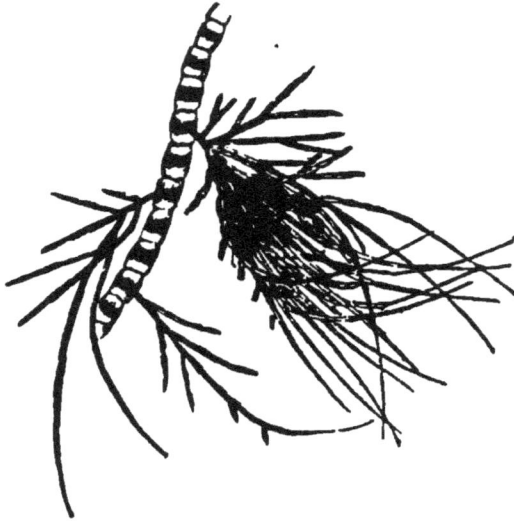

Fig. 90.—Draparnáldia glomeráta.

9. BULBOCHÆTE (Fig. 91).

This genus can always be known by the swollen or bulbous bases of the long hairs that tip many of the .cells. It grows on larger Algæ, or on the leaflets of Ceratophyllum or other aquatic plants (Fig. 91).

Fig. 91.—Bulbochæte.

CHAPTER IV.

RHÍZOPODS.

THE Rhizopods are the lowest animals in the scale of life. Scarcely more than a drop of jelly-like protoplasm, the lowest of these lowly creatures live, move, eat, and multiply. Some are so far down in the scale that they are actually only a particle of soft and unprotected protoplasm, moving, like the common *Amœba*, which is one of the Rhizopods, by protruding long, thread-like projections of its own substance from any part of its body, and withdrawing them again into its substance, where they entirely disappear. These protruded parts, by means of which the creatures move and capture their food, are called pseudopodia, from two Greek words, meaning false feet. And since they often extend to long distances from the body of the animal, dividing and branching somewhat after the manner of roots, the group of lowly animals producing these pseudopodia is named the Rhizopods, or root-footed, a word also from the Greek.

The Amœba, and those Rhizopods nearest to it in structure, are formed of naked protoplasm; they are simply a drop of living jelly. But some higher in the same group secrete or build around their soft bodies a protective shell, often of exquisite form and remarkable construction. Thus the members of one genus, *Difflú-gia*, build themselves shells of sand grains cemented together with the most perfect regularity, every grain exactly fitting to its place. Yet, when the young Diffúgia happens to be where suitable sand is scarce, it will build its shell of diatoms, often using those that are longer than the completed covering, attaching them lengthwise, side by side, and parallel to each other. Another genus, *Arcélla*, secretes from its body a brown shell of delicate membrane which, with a high power, is seen to be formed in minute hexagons. And still another, *Clathrulina*, the most beautiful of all the fresh-water Rhizopods, lifts itself on a long stem, and there surrounds its body by a hollow latticed sphere, and through the openings in the walls extends its pseudopodal rays in search of food.

In the unprotected forms—those without a shell—the pseudopodia are protruded from any part of the body; in those preparing shells they are protruded from that portion of the body immediately in contact with the mouth of the shell, through which they often extend for a long distance as very fine, branching threads. With a few exceptions the bodies of the Rhizopods are colorless; in those exceptions the coloration is usually due to the presence of colored food, and so is diffused throughout the entire protoplasm, or it is confined to the parts near the surface, the central portion being nearly colorless. The pseudopodia are never colored.

Not only do the Rhizopods move by means of these

"false feet," but they capture food with them, consuming both plants and animals. Diatoms, Desmids, Infusoria (Chapter V.), Rotifers (Chapter VIII.), almost any living thing small enough to be seized, is acceptable. When a desirable morsel is found, the end of the pseudopodium touching it usually expands, and a wave of the body substance flows along it until the object is surrounded, like an island of food in a sea of protoplasm. The whole broadened pseudopodium is then withdrawn into the body, carrying the food with it; or, if the captured object is unusually large, or if it struggles a good deal, several pseudopodia may come to the assistance of the first, or a great wave-like outflow from the body may envelop both pseudopodia and food.

These curious animals have no distinct mouth and no distinct stomach. The mouth in the shell-less ones is formed at any point on the surface wherever the creature chooses to open itself and take in the food particle; and the stomach is in any part of the internal substance; the food is digested wherever it may happen to enter and remain. They have no eyes, yet they seem to direct their course and avoid unpleasant or injurious obstacles. They have no nerves, yet when disturbed they contract into a small ball-like mass, or withdraw themselves into their shell. They also appear to feel some sort of sensation of hunger, for they are often seen to take food, and they select what they like.

They are very numerous and common. They are to be found in any shallow pond, or pool, or body of still water. They glide among aquatic plants and Algæ, especially on the lower surface of water-lily leaves, and among Myriophyllum and Ceratophyllum. Sphagnum moss is sure to contain them in abundance, as has already been stated on page 61. But the mud is an accessible and fruitful source of supply. To obtain them, gently scrape with a big iron spoon or the edge of a tin dipper the surface of the ooze from the mud in shallow ponds, and transfer it to the collecting-bottle. Let the muddy mixture stand for a few minutes until the Rhizopods settle towards the bottom, and carefully pour off some of the water, adding more ooze if desired. Pour the mud and water into saucers, and set them near the window, when the Rhizopods will make their way to the surface, and may be removed by the dipping-tube. Do not place the saucers in the sunlight; Rhizopods prefer a little shade. They are invisible, consequently the collector must collect on faith, as he must usually do when out on a microscopical fishing tour. But he will seldom be disappointed if he gathers the surface ooze from the edges of somewhat shady ponds, and avoids those places long exposed to the sun, and never sinks the dipper into the thick black mud, which contains no animal life of any kind.

They are small and easily overlooked in the field of the microscope, but when one of the unprotected forms and a single shell-bearing Rhizopod is recognized, the beginner will never again overlook any of them in the material on his slide. The Amœba will probably be the first seen, as a colorless, jelly-like body, very soft, and changeable in shape, slowly moving forward and suddenly altering its course and extending itself in numerous long, blunt, finger-like pseudopodia, lengthening or shortening at the creature's will. Or he may see a small pear-shaped collection of sand-grains slowly moving about the slide, apparently without a cause, but a careful examination of the narrow or stem end of the pear will show the long, fine, and colorless pseudopodia issuing from the mouth, and he will know it to be a Rhizopod. After he has recognized a living shell he will have no trouble thereafter in knowing a dead one, and by referring to the following Key he will be able to learn its name, unless it is a very uncommon species.

Key to Genera of Rhizopods.

1. Body without a shell (*a*).
2. Body with a shell (*e*).

   a. Without fine, hair-like rays; pseudopodia thick and blunt (*b*).

   a. With fine, hair-like rays on all parts of the body (*c*).

   b. Body colorless, very changeable in shape. *Amœba*, 1.

   c. Body orange or brick-red, with pin-like rays. *Vampyrélla*, 2.

   c. Body colorless or greenish (*d*).

   d. Rays stiff, forked at the ends; body often green. *Acanthocýstis*, 3.

6*

d. Rays flexible, not forked.   *Actinóphrys*, 4, or *Actinosphœrium*, 5.

e. Shell formed apparently of sand-grains (*f*).

e. Shell not formed of sand-grains (*g*).

e. Shell a latticed globe on a long stem.   *Clathrulína*, 12.

f. Not inclined; pear-shaped, or globular with spines at the summit.   *Difflúgia*, 6.

f. Inclined; circular or oblong, thicker and with spines at the rear.   *Centropýxis*, 7.

g. Shell brown (*h*).

g. Shell colorless, ovoid, not curved (*i*).

g. Shell often yellowish, ovoid, curved (retort shaped), mouth circular.   *Cyphodéria*, 11.

h. Circular, with or without marginal teeth.   *Arcélla*, 8.

i. Mouth smooth, circular; shell inclined, without spines.   *Trinéma*, 9.

i. Mouth serrated; shell not inclined, formed of hexagonal plates; often spinous.   *Eúglypha*, 10.

1. Amœba (Fig. 92).

There is hardly a living animal so soft and changeable in shape as this. It may not retain the same form for a second at a time. The soft body protrudes thick, blunt, finger-like pseudopodia from any part of its surface, but usually from the front margin, or that edge at the forward part of the moving creature. The front may, with scarcely a warning, become the rear as the animal changes its course, by emitting pseudopodia from some other portion, travelling off in the direction towards which they extend. The semi-fluid contents of the body are colorless, unless tinged by the food or by the presence of numerous dark particles. The movements are sometimes quite rapid, the Amœba extending its pseudopodia, keeping them extended in advance, and gliding along as though the body were formed of the white of egg. In the figure it is shown with many short pseudopodia, as it often appears immediately after it is placed on the slide, and before it has learned where it is, and has prepared to move in some definite direction. The posterior extremity, when the Amœba is in motion, may be entirely smooth, or it may show a cluster of very short pseudopodia, giving it a velvety or mulberry appearance. Suddenly a blunt, thick finger projects from the part, and Amœba at once reverses its course, the pseudopodia at the front being withdrawn, and disappearing in the substance of the body. The observer can never predict what an Amœba will do next. It is very common in the ooze of shallow ponds and on the leaves of many aquatic plants. Its body usually contains a number of diatoms, which form part of its favorite food, and it is a strange fact that the food is usually taken by what, at the time, is the posterior extremity. There are several species.

1. Body large, colorless or blackish; pseudopodia finger-like, blunt. *Amœba próteus*, Fig. 92.
2. Body small, colorless, rather sluggish; often floating

freely, and star-shaped, with several conical, acute,
straight, or curved pseudopodia
radiating from the spherical central body. The form changes
very slowly. *A. radiósa.*

Fig. 92. — Amœba pró-
teus.

3. Body irregular in shape; pseudopodia usually few, short, thick,
   and directed forward; posterior
   portion of the body with a vil-
   lous or velvet-like patch of very short, colorless
   pseudopodia. *A. villósa.*

2. VAMPYRÉLLA LATERÍTIA (Fig. 93).

A red or orange colored, Amœba-like creature with
this name is not uncommonly found in early spring
among thick growths of Spirogyra, for which it has a
special fondness. It does not very quickly nor frequently change its shape, yet its movements are quite rapid.
Its pseudopodia are colorless and transparent, being

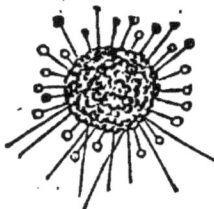

Fig. 93.—Vampyrélla
lateritia.

formed by a short outward flow of the
colorless central body substance, the red
color being confined chiefly to the surface. It also has short, fine rays like
threads, and many pin-like projections,
by which, in connection with its color,
Vampyrélla may be easily recognized.

These pin-like rays consist of a short, fine stem with a
little bulb on the end, so that each looks very much
like a pin with a big head. They may appear on all

parts of the body, but usually they are on the rear end only when the animal is moving. They often appear very suddenly, and as quickly disappear.

Vampyrella's favorite food seems to be the cell contents of Spirogyræ. It selects a fresh and healthy plant, and settling down upon it, proceeds to perforate the cell-wall, and to remove the color bands with the entire cell contents by drawing them into its body, leaving the cell quite empty, with a ragged hole in the side. I have seen one Vampyrella remove the contents from seven Spirogyra cells in succession before its appetite was satisfied.

3. ACANTHOCÝSTIS CHÆTÓPHORA (Fig. 94).

Body spherical, soft, usually colored green by the numerous green granules within. When the animal changes its shape, which it seldom does, it only becomes oval or slightly irregular in outline. The pseudopodia are very fine and hair-like, springing from all parts of the surface, but the peculiarity by which it may easily be known is the dense growth of spines covering the entire body, their ends being forked or divided into two short, straight, diverging branches. To see these forked ends demands a rather high-power objective, as they are small, but the spines themselves are apparent to a comparatively low-power. They seem not very securely fastened to the animal; some of them quite often become loosened and drop off, especially if the Rhizopod is not in a healthy condition.

When food is to be taken into the body, a part of the

surface with the adherent spines is lifted up, carrying the spines to one side, and a wave of protoplasm, the body substance, flows out to receive and surround the food brought down by the pseudopodia. It is drawn into the body, the surface closes, and the spines again cover the spot. This may happen at any part of the surface.

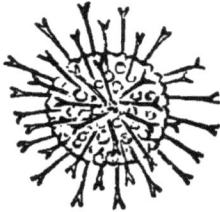

Fig. 94.—Acanthocystis chætophora.

Acanthocystis is often found among the leaflets of Myriophyllum, the roots of Lemna, or floating freely in quiet water. It is rarely found in the mud.

4. ACTINÓPHRYS SOL (Fig. 95).

This is one of the commonest of aquatic microscopic animals. It may be found floating in every quiet pond or pool, or swimming among the leaflets of nearly every gathering of water-weeds. Its body is usually colorless and almost transparent, seeming to be formed of a collection of small bubbles, so that it has a foamy appearance. It bristles with numerous long, fine rays springing from the whole surface. It moves in a slow, gliding way that has not been satisfactorily explained, but which can hardly be produced by the hair-like rays, for they are motionless, and apparently used only for capturing food. Yet it slowly floats across the field of view, seldom changing its shape; or it remains suspended almost stationary in the water with all its rays extended, and so resembling the pictures of the sun in an almanac

that it has received the name of the "sun animalcule." The rays are seldom entirely withdrawn.

It feeds on smaller animals and the spores of Algæ. When an animalcule comes in contact with the rays it seems to lose some of its power of motion. It appears to become partially paralyzed, gliding

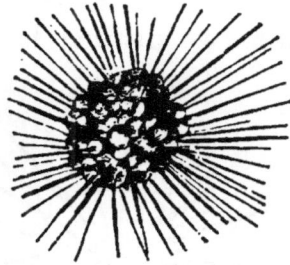

Fig. 95.—Actinóphrys sol.

down the ray, often surrounded by a small drop of pro-
toplasm, until it nears the body, when a larger wave flows out and receives it. The little masses of digest-
ing food can be seen inside the body, where the green coloring usually turns to brown.

5. ACTINOSPHÆRIUM EICHHÓRNII (Fig. 96).

At first the beginner will confound this Rhizopod with Actinóphrys sol, which it resembles in appearance when seen with a low-power objective. It is larger than the "sun animalcule," but this is a distinction of no value unless the observer has happened to find Actinophrys first, and to have become familiar with its appearance and structure. In *Actinosphærium*, however, the ray-
like pseudopodia are quite large and coarse, and they ta-
per to their free end from a thickened base at the sur-
face of the body. The body itself, as the student will notice if he uses a ¼ or ⅛ inch objective, is formed of an external layer of large vesicles or bubbles, and a central mass of smaller bubbles. In this bubble-like

structure it also resembles Actinophrys, but it seems
less like a drop of froth, for the bubbles are larger, and
the two distinct layers of two different sizes at once
show that the Rhizopod is
Actinosphærium. But there
is another and more important difference, which the beginner will not observe unless he searches for it with
a high-power ($\frac{1}{4}$ or $\frac{1}{8}$) objective. Each ray has a thread
or fine rod running length-

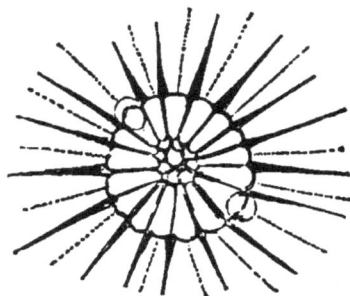

Fig. 96.—Actinosphærium Eichhörnii.

wise through its middle, and differing slightly in color
from the softer part of the ray. This rod begins within
the body below the outer layer of larger bubbles, passing between them and extending almost to the end of
the pseudopodal rays, which are seldom entirely withdrawn into the body.

Actinosphærium is sluggish, moving slowly and often
remaining motionless for a long time in one spot. It is
frequently found in company with Actinophrys, among
Lemna and other aquatic plants.

It feeds on other animals as well as plants, taking
larger victims than the "sun animalcule." The Rotifers (Chapter VIII.) seem its favorite food. A free
swimming animalcule or Rotifer coming in contact with
the long rays seems, as with Actinophrys, to become in-
capable of escape; it is then slowly drawn into the body
and digested.

6. DIFFLÙGIA (Figs. 97, 98).

Shell brown, pear-shaped, ovoid or nearly spherical, and formed of angular sand-grains cemented together. The upper part, the summit, may be rounded, and roughened only by the edges of the sand-grains, or rounded and bearing several pointed spines also formed of sand. The lower part may be prolonged as a short neck, at the end of which is the mouth for the passage of the pseudopodia, or the shell may have no part resembling a neck. The animal which builds this protective case lives inside of it, and is a little mass of colorless, or sometimes greenish, protoplasm, somewhat resembling an Amœba, and almost entirely filling the cavity of the shell. The mouth is circular, and may be either smooth or with several rounded teeth or lobes on its inner edge. No part of the animal in any of the shell-bearing forms, except the pseudopodia, ever passes through the mouth. When the shell is made the animal never leaves it, unless it is broken by the cover-glass; then it will at times creep out and die.

The pseudopodia are blunt and colorless. They drag the shell about with the mouth downward, and capture food as in the naked Rhizopods. When they are withdrawn, the shell appears like a dead thing, and may roll about the slide at the will of the observer or the mercy of the currents. But often while the student is looking at an apparently dead shell of sand, a blunt little colorless wave issues from the mouth, lengthens and narrows, is followed by another and another, until the shell is raised and moved slowly away.

There are several species of the genus *Difflúgia*, of which the following are about the commonest. They are found abundantly in the mud and among Sphagnum.

1. Shell pear-shaped (Fig. 97), without spines, although the summit may be prolonged into one or two points; usually formed of sand-grains, sometimes with adherent diatoms; occasionally formed entirely of diatoms; mouth at the narrow end, circular, smooth, without teeth or lobes. The body within the shell is usually green, sometimes colorless; pseudopodia colorless, thick, blunt. It is almost as fond of the cell contents of Spirogyra as is Vampyrella, and obtains them in a similar way; but instead of appearing to suck them out of the cell, *Difflúgia pyrifórmis* pierces the wall, inserts its pseudopodia, with them surrounding the color bands and other cell contents, lifts the whole out and passes it into the body within the shell. I have seen a single Difflugia empty four Spirogyra cells in succession. This species is common. *Difflúgia pyrifórmis*, Fig. 97.

Fig. 97.—Difflúgia pyrifórmis.

2. Shell nearly spherical, with from one to twelve, usually three or seven, pointed spines arranged in a circle around the upper part, and formed of sand-grains. These spines are hollow, and communicate

with the cavity of the shell, but the animal probably builds them for ornament, as it does not seem
to use them. The mouth of the shell occupies the
end opposite to the spine-bearing summit, and
when the shell is turned over so that this opening
is directed upward, it will be seen to be lobed or
scalloped, the lobes varying from six to sixteen,
being usually about twelve. They may in some
forms be rather sharp-pointed, almost like short
teeth. They are directed towards each other across
the opening. It is a difficult matter to get the
shell in such a position that the
observer can look down into the
mouth, but it may sometimes be
done by tapping the cover-glass with
a needle so as to roll the Rhizopod
about, and occasionally, by one of
those lucky accidents that sometimes occur, it places itself in good
position. The soft body is colorless or brownish, and
the pseudopodia are thick, blunt, and numerous. The
species is common in the ooze. _D. coróna_, Fig. 98.

Fig. 98.—Difflúgia coróna.

3. Shell spherical, without spines; mouth circular,
smooth, without lobes or teeth. This species is
found with the preceding. _D. globulósa._

4. Shell long and narrowly pear-shaped, the summit
prolonged into a central sharp point; mouth circular, smooth, without teeth or lobes. Common. _D.
acumináta._

7. CENTROPÝXIS ACULEÁTA (Fig. 99).

The shell of this Rhizopod is usually formed of sand-grains, and is brown in color, but sometimes it consists of a brown membrane with scattered adherent sand-grains. I have also met with shells formed entirely of small diatoms fitted together as beautifully and accurately as the sand-grains of Difflugia. These diatom shells were found in an aquarium, and were probably built of these plants because suitable sand was not to be had. *Centropyxis*, when seen in side view, appears as if it had once been a hemisphere with the mouth near one side of the flat surface, but that while it was soft the convex part had in some way been pushed over towards one side, thus leaving the shell oblique or inclined, the back part being much thicker than the front, the upper surface sloping down from the deeper rear to the thin front margin, the circular or oval mouth remaining nearer the thin border. The figure shows the under part of a shell, which, in this position, appears almost circular. The spines on the thick part are usually sharp-pointed, and vary in number from one to nine. The body of the animal is colorless, and the pseudopodia are blunt and finger-like. This is the only known species. Common.

Fig. 99.—Centropýxis aculeáta.

8. ARCÉLLA (Figs. 100, 101).

When seen from above or below, the shell of *Arcélla*

seems like a disk with a pale circular spot in the middle. When seen in side view it has a flat lower surface and a more or less strongly convex or elevated upper surface. In color it is usually some shade of brown, but may be almost black. In very young specimens the shell is often nearly colorless. It is generally transparent. The mouth of the shell, in the centre of the flat surface, is circular and smooth. The body of the animal is colorless, and is attached to its home by fine threads of its own substance. There are several species, recognizable by the form of the shell.

1. Margin of the shell smoothly circular. Common everywhere. *Arcélla vulgáris*, Fig. 100.

2. Margin of the shell with several teeth, so that it resembles, when seen from above or below, a wheel with pointed cogs. Not as common as the preceding. *A. dentáta*, Fig. 101.

3. Shell somewhat balloon-shaped when seen in side view; higher than wide, the sides often depressed in wide facets. Not rare. *A. mitráta*.

Fig. 100.—Arcélla vulgáris.

Fig. 101.—Arcélla dentáta.

Fig. 102.—Triuéma éuchelys.

9. TRINÉMA ÉNCHELYS (Fig. 102).

This shell is pouch-shaped, colorless, small, and in-

clined, so that when in motion with the mouth downward against the slide the rounded summit is lifted
obliquely upward.   It is somewhat narrower at the
lower part, and the mouth is a short distance within
the shell, the front or lower edges seeming to curve inward to meet it.   The body of the animal is colorless.
The pseudopodia are very fine, thread-like, and few in
number.   The Rhizopod is common everywhere in wet
places; it is also one of the smallest, and the shell is
often found dead and empty.   The figure shows it in
side view.   The aperture of the shell is seen to be beaded when examined with a high-power objective.

10. EÚGLYPHA (Fig. 103).

The shell of *Euglypha* is ovoid, colorless, and transparent.   Under a high power it is seen to be composed
of many oval or hexagonal plates arranged in rows,
those towards the widest part of the shell overlapping
those in front.   The mouth is circular or oval, but the
projecting points of the plates give it a toothed, saw-
like edge.   There are several species, but they all may
be known as Euglyphæ by this serrated or saw-toothed
mouth.   The upper part and the borders of the shell
are either with or without spines, or they may bear fine
hairs.   The animal itself is colorless, and almost entirely fills the cavity of the shell, to which it is attached,
apparently by the summit only.   The pseudopodia are
very delicate and often branched.   The animal moves,
like all the shell-bearing forms, with the mouth of the

shell against the slide or other object over which it creeps.

1. Shell without spines, or with four or six near the summit and arranged in a circle at equal distances apart, pointing upward and varying somewhat in length. Quite common in the ooze of ponds. *Eúglypha alveoláta*, Fig. 103.

Fig. 103.—Eúglypha alveoláta.

2. Shell with a cluster of spreading spines springing from the centre of the summit. Common in Sphagnum. *E. cristáta.*

3. Shell with the summit and sides fringed with bristles. Common in Sphagnum. *E. ciliáta.*

11. CYPHODÉRIA AMPÚLLA (Fig. 104).

Shell yellowish, or sometimes colorless, shaped like a chemist's retort, the mouth being at the narrow, curved end. The summit is rounded, sometimes with a central point or small knob. The shell, when highly magnified, is seen to be formed of minute hexagons. The animal is, as usual, colorless, and nearly fills the semitransparent case. The pseudopodia are numerous and often forked. When moving, the mouth of the shell is in contact with the object over which the Rhizopod is travelling, and the body of the shell is held obliquely upward or almost parallel with the slide. The figure shows an empty shell. There is but one

Fig. 104.— Cyphodéria ampúlla.

species, which is quite frequent in the ooze of ditches and ponds.

12. CLATHRULÍNA ÉLEGANS (Fig. 105).

A hollow globe of silicious lattice-work elevated on a crystalline stem. Within this exquisite dwelling the spherical, colorless animal lives, extending its fine long pseudopodal rays through the almost circular windows in search of food. The stem is attached to aquatic plants or other submerged objects. *Clathrulina* is the only fresh-water Rhizopod that is not free-swimming. It is common in many ponds, attached to the rootlets of Lemna.

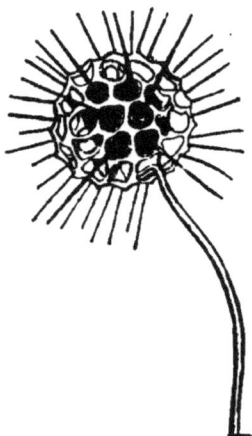

Fig. 105. — Clathrulina élegans.

In this small book it is only possible to refer to a very few of the commonest of these beautiful and interesting animals, about whose life history very little is known. They form a department in which there is room for much original investigation. Those who desire to pursue the subject, or to know more of the Rhizopods than can be included here, would do well to refer to Dr. Leidy's "Fresh-water Rhizopods of North America," published by the United States Geological Survey of the Territories, or to Mr. Romyn Hitchcock's "Synopsis of the Fresh-water Rhizopods," a useful condensation of Dr. Leidy's splendid work.

CHAPTER V.

INFUSÓRIA.

The reader probably knows the Infusória under the name of animalcules, a word only meaning small animals, which the Infusoria certainly are. But a mouse is also a small animal; so is a Water-flea (Chapter X.) and a Rotifer (Chapter VIII.). Infusorium for a single one of a group of certain microscopic creatures, and Infusoria as the plural, are better words than animalcule, with no danger of conveying an incorrect meaning. The Infusoria were so named because they were first discovered in infusions, that is, water in which animal or vegetable substances had been soaking and decaying. Since that time the creatures have been obtained in great abundance and variety in even the sweetest of fresh waters, although they abound in astonishing numbers in many infusions. The beginner has only to place a handful of hay in a tumbler of water, and allow it to soak for a week or two, when he will have as many Infusoria as he may want for examination. They are also plentiful in every ditch and pool of still water. No collection of Algæ, aquatic plants, or Rhizopods, can be made without, at the same time, gathering very many Infusoria.

One of the best ways to collect the little creatures is to gather aquatic plants and Algæ without taking them

7

from the water. If the plants among which they conceal themselves and search for food are lifted out of the
pond, the water running off washes away all the animals
you are seeking. So take the water in the dipper, or
float the plants into the bottle, which should never be
entirely filled, nor corked for any length of time. The
Infusoria are very fond of fresh air; they rapidly ex-
haust the oxygen in solution in the water, dying quickly,
and going to pieces almost as soon as dead. Give them
plenty of air in the collecting-bottle, and at home pour
the gathering into a broad dish so as to have a wide surface exposed to the atmosphere. The plants as well as
the Infusoria do better in such quarters. They are also
usually fond of the light, and will soon make their way
to the side of the vessel nearest the window, and the
dipping-tube put in at that side will often capture creatures that avoid the shadier parts. To obtain those that
are free-swimming, that is, those that are never permanently adherent to the leaflets of plants nor the filaments of Algæ, as many of the most interesting are,
they can be transferred to the slide by the dipping-tube,
and the drop covered by the thin glass, when they are
ready for study. Those attached to plants can be found
only by cutting off a small piece of Myriophyllum or
other water-weed and examining it under the microscope. In these cases it is necessary to lift the piece of
weed from the water, but it can be moved gently, and
at once placed in a drop ready for it on the slide. Some
of the most interesting kinds of Infusoria are found ad-

herent to Ceratophyllum and other plants with finely divided leaves. Every part should be searched with the microscope, especially the angles between the leaflets.

The bodies of the Infusoria are usually very soft and delicate. Some of them are so flexible that they can double and twist themselves almost as well as a worm. Others are hard, and some are even covered by a transparent case secreted from their own body. This case is called a *lorica*, and is used as a shelter for the soft and otherwise defenceless animal. When frightened it quickly withdraws itself to the bottom of the lorica, and remains there in a little, almost shapeless, heap, until the danger is past. Then it slowly rises up to the front of the lorica, protrudes the front part of the body, opens the organs by which it creates currents in the water, and so fishes for the food those currents bring to its mouth. These loricæ are usually permanently attached to plants or other submerged objects. They are also generally transparent and colorless, but sometimes, as they become old, the color changes to a rich, translucent chestnut brown. In other Infusoria the loricæ are not hard and transparent, but soft and delicate. These are usually made of innumerable little particles of dirt fastened together by a sticky substance secreted from the animal's body. Almost any small particles floating about and striking against the sticky mass will be quite sure to adhere, and so help build up the soft sheath that serves the Infusorium as a protective covering, and sometimes effectually conceals it from

the microscopist who may be seeking it.   But they are
not formed entirely by accident.   They are built chiefly
of·those little particles brought to the animal by the
currents produced by the organs it has for that purpose.
These currents contain the food which the Infusorium
cannot go to seek as the free-swimming kinds can do,
for the loricæ building animals are almost as permanently fastened to their loricæ as is a snail to its shell.
Sometimes the Infusorium will leave its lorica when
the water has lost most of its oxygen, and the poor thing
is nearly smothered, and it leaves only to die.   But it
generally prefers to die at home, for when the time
comes the little creature retires to the bottom of the
lorica, contracts into a heap, and quietly goes to pieces.

There are also some that form loricæ and are still
free-swimming, carrying the house about with them.
They also retire to the rear when frightened, and some
even have a little piece of hard substance on the front
of the body with which they plug up the entrance, and
so make all secure.

There are others that form a stem and branches like
the trunk and limbs of miniature trees, the colorless
animals being fastened to the ends like so many leaves.
In some of these the animals can contract themselves
into little balls when frightened ; in others the branches
contract into coils and pull the animals away from harm ;
in still others the whole tree-like colony, stem, branches,
and animals, contract and pull themselves down against
the plant to which the stem is attached.   And in still

others, the *Vorticéllæ*, there is but one stem with a single bell-shaped body on the end, but the stem contracts into close spirals and suddenly draws the animal down. When the danger is past, the stem slowly uncoils, the branches spread themselves, the animals expand, and all is as before. Indeed, the variety of form and habit in the Infusoria is almost infinitely great.

The general opinion is that "animalcules" have no color. This is a mistake. The majority are almost colorless, but green, crimson, yellow, indigo blue or almost black Infusoria are not uncommon, and the loricæ, as stated, often become brown.

The free-swimming Infusoria are more abundant than the attached ones, and much more difficult to examine because they will never stand still. But how do these creatures, all of which are invisible without the microscope—how do they move? For this purpose they have organs of two kinds, and they are separated into two great classes according as they possess the one kind or the other. In some there are one or more long, colorless lashes which extend from the front of the body, beat against the water, and so row the animal about very rapidly. Each of these lashes is called a *flagellum* (plural, *flagella*). In others there are on the body short, very fine hairs, which are continually vibrating so rapidly that they are often invisible even under a high-power objective. The short hairs are called cilia, and it is their action on the water that urges the animal about, even more quickly than the flagella. The cilia may be confined to a circle around one end, or they may be on the lower surface only, or the whole body may be covered with them. Infusoria with cilia are more numerous than Infusoria with flagella. They are, however, not the only ciliated animals. The Rotifers are well supplied, and certain small aquatic worms have the entire body ciliated.

Although the Infusoria are so abundant that scarcely a drop from any pond or ditch can be examined without exhibiting some, the beginner will, I fear, have trouble in studying them, they are so lively and so small. The stage must be kept in continuous motion to counteract the motions of the Infusorium and keep it in the field, so it can be seen as anything more than a whirling speck, and high-powers are needed to examine it. But the beginner's object will be gained if he learns to know an Infusorium when he sees one, and if he learns the names of some of the largest and commonest. Many can be seen with a one-inch objective, but to ascertain whether any special one has cilia or flagella will demand a one-fifth inch or higher power lens, and without knowing this the Infusorium cannot be identified. But "it is only the first step that costs." Any work or study is always hardest at the beginning. When the student has identified one Infusorium he will have little trouble with what comes after. The attached forms will not be very difficult even at the first, if a sufficient magnifying power is used, for since they are fastened by stem or lorica to another object, they can be examined at leisure.

None of these creatures can be preserved as permanently mounted objects. Many chemical solutions and mixtures have been recommended for killing and keeping them, but none are satisfactory, the soft bodies going to pieces and melting away almost as soon as after a natural death. If the beginner is very much annoyed by the incessant movements of the free-swimming kinds, and he desires to see how they look when quiet for a moment, the following solution will help. It answers the purpose well in some cases, while in others it is worthless. It always kills, but does not always preserve after death. It is used by allowing a small drop to run under the cover-glass and to mingle with the drop of water containing the Infusoria. Any druggist can make it, but caution him to use not more than half a drachm (half a teaspoonful) of water, or you will be terrified by his bill. If this small quantity is made it is not expensive.

To the half drachm of water add as much iodide of potassium as it can be made to dissolve, and to this solution add as much iodine as the solution can be forced to dissolve. This ends the druggist's part. It only remains for you to add enough of the mixture to clean water to make the color a rather deep amber. The proper strength can be learned by experiment. If it kills, and then destroys too quickly, add more water; if it does not kill quickly enough, drop in a little more of the iodine mixture.

A weak solution in water of the *per*-chloride of iron

has also been recommended for this purpose, but its action is similar to that of the iodine solution, and not more satisfactory.

The following Key refers to only a few of the commonest Infusoria in fresh water and vegetable infusions. To include a tithe of those most frequently seen in such places is an impossibility. When the beginner learns that there are fifty known species of Vorticélla alone, and about thirty of Mónad, he will see that it is possible to refer in the most superficial way to only a very few of these abundant and attractive creatures.

Key to some Genera of Infusoria.

1. Free-swimming (*f*).
2. Not free-swimming; singly or in clusters on a stem (*a*).
3. Not free-swimming; in a transparent or granular lorica (*b*).

   a. Stem much branched, neither it nor the animals contractile. *Dendrómonas*, 1.

   a. Stem much branched, both it and the animals contractile. *Carchésium*, 2.

   a. Stem much branched, only the animals contractile. *Epistýlis*, 3.

   a. Stem not branched, contracting into spirals. *Vorticélla*, 4.

   b. Loricæ vase-shaped, transparent (*c*).

   b. Loricæ soft, granular, brownish (*e*).

   c. Attached to each other to form colonies. *Dinóbryon*, 5.

c. Not attached to each other (*d*).

d. Lorica without a stem, adherent by the narrow base. *Vaginicola*, 6.

d. Lorica without a stem, adherent by the broad side. *Platýcola*, 7.

d. Lorica with a short stem. *Cothúrnia*, 8.

e. Extended animal trumpet-shaped. *Sténtor*, 9.

f. With one or more flagella at the front (*g*).

f. Without flagella, but with cilia (*h*).

g. Body very changeable in shape, colorless. *Astásia*, 10.

g. Body very changeable in shape, green or red. *Eugléna*, 11.

g. Body not changeable in shape, colorless, notched in front. *Chilómonas*, 12.

g. Body not changeable in shape, green, with a short, stiff, colorless tail. *Phácus*, 13.

g. Body not changeable in shape, green, united in a revolving colony. *Uvélla*, 14.

h. Cilia on the entire surface (*i*).

h. Cilia confined to the lower or flat surface (*k*).

i. Neck long, very elastic and extensile. *Trachelocérca*, 15.

i. Neck long, flattened, not extensile. *Amphiléptus*, 16.

i. Body brownish, slipper-shaped. *Paramœcium*, 17.

i. Body green, red, blue or almost black; ovoid or trumpet-shaped; cilia largest on the front. *Sténtor*, 9.

7*

k. Cilia large, few, scattered (*l*).

k. Cilia fine, numerous (*m*).

l. Body more or less circular in outline. *Euplótes,* 18.

l. Body more or less oblong in outline. *Stylonýchia,* 19.

m. Mouth followed by a conical tube of rods. *Chílodon,* 20.

m. Mouth followed by a brown, sickle-shaped membrane. *Loxódes,* 21.

1. DENDRÓMONAS (Fig. 106).

The stem is many times divided into numerous branches, and the branches themselves are also much divided, with one small Infusorium at the end of each.

Fig. 106.—Dendrómonas.

The whole has a beautiful but colorless tree-like appearance, the stem being often found attached to Ceratophyllum. The animals have each two flagella, but they are visible only to a high-power objective. There is no special mouth. A particle of food dashed down by the flagella against any part of the body sinks into its soft side and is thus swallowed without a throat. The whole colony is often more branched than is shown in the figure. It can be recognized with a good one-inch objective.

2. CARCHÉSIUM (Fig. 107).

The stem, attached to plants or other submerged

objects, is divided at the summit into many branches, with one Infusorium at the end of each, and many others scattered along them with shorter branches of their own. Through the main stem and through all the branches there extends a cord-like muscular thread that suddenly contracts when the animals are frightened or disturbed, and pulls the whole colony down towards the point of attachment to the plant. But the branches may contract one at a time and draw their burden of Infusorial fruit down to the main stem without disturbing any other portion of the colony, or all the branches may contract at once. Therefore, while the animals on the branches are

Fig. 107.—Carchésium.

connected together, they are still somewhat independent. The front border of each body is surrounded by a circle of cilia visible under a high power. They are the only cilia on the body. When the animal is contracted they are folded together, each body then resembling a little ball. They vibrate rapidly, producing circular currents that bring to the mouth any food-particles that may be in the vicinity. The entire colony is colorless, and may include as many as a hundred Infusoria on the branches. It can be seen by a low-power objective. The independent contraction of the branches and the stem will distinguish it from all other tree-like Infusoria.

3. Epistýlis (Fig. 108).

As in the two preceding, the stem of *Epistýlis* is also often much branched. The Infusoria at the ends of the branches can alone contract, which they often do with a jerk, settling back as if they meant to impale themselves, or dropping and nodding like flowers fading on their stems. The bodies of the expanded animals are somewhat bell-shaped, their widest part being the free end which closes when the body contracts.

The front border is encircled by a row of cilia, to be properly discerned only by a high-power objective. The one-inch glass, however, will show the rapid currents produced, because all small particles in their neighborhood are caught up and dashed around in the mimic whirlpools. The animals select from these streams anything they may want and let the rest sweep by.

Fig. 108.—Epistýlis.

They have a distinct mouth near the centre of the front part. The entire colony is usually colorless. It is often attached to Ceratophyllum.

4. Vorticélla (Fig. 109).

The unbranched stem of *Vorticélla* contains a zigzag muscular thread like a thin cord, which contracts into close coils very suddenly, and draws the Infusorium down with it. The Vorticellæ are very common, scarcely a leaflet of any aquatic plant is without them. They are usually colorless, although green ones do occur. The body is bell-shaped, the narrow part of the bell being fastened to the top of the stem. The front border is surrounded by a circle of fine cilia which need a high power to show them. They produce currents in the water similar to those of Epistylis, and for the same food-collecting purposes.

The contractions are surprising in their suddenness. While the observer is quietly gazing at the graceful creature whirling its cilia and making tremendous whirlpools on a small scale, it disappears like a flash, and the student feels like looking for it on the table. But presently it slowly begins to rise from the plant against which it was crouching, and the coiled stem lengthens as it straightens.

Fig. 109.—Vorticélla.

Very often it hardly extends before it again leaps out of sight, or close to the object supporting the stem. When the stem throws itself into spirals, the body of the animal folds together into a ball.

This will probably be one of the first Infusoria to attract the beginner's attention, and he will think it a wonderful thing, as it is. The figure shows some extended and some contracted. They are often found in clusters, sometimes of a hundred or more, all bobbing and swaying in a very curious way, for when one contracts it usually sets them all off.

5. Dinóbryon (Fig. 110).

In the early spring, as early as March, among the Algæ then found so abundantly in the shallow pools, colonies of very small, vase-shaped loricæ are often obtained. They are sometimes attached to a plant or filament of alga, or as often they float freely through the water, being fastened to the plant by a very slight hold. The loricæ are transparent and colorless, and may be overlooked, but the Infusorium within each one is rather conspicuous to even a low-power objective, for it has a narrow green band on each side of the body, and often a minute red eye-like spot in the centre of the front border. The loricæ are united together by one or two being attached to the front edge of the one behind them, until branching colonies of some size are formed. The front border of each enclosed Infusorium bears two flagella, one long and one short, but they are seen with difficulty even with a moderately high-power objective. The lashing of all the flagella in a large colony urges it quite rapidly through the water. According to my experience *Dinóbryon* is seldom found in the summer.

Fig. 110.—Dinóbryon.

6. Vaginícola (Fig. 111).

The lorica is colorless, transparent, and about three times as long as broad. In form it is long, vase-shaped, or nearly cylindrical, the base, or the part fastened to the plant or other object, being usually rounded. The animal, when it projects, extends for a considerable distance beyond the opening at the front of the lorica. When frightened, or disturbed in any way, it quickly closes up its broader front part, and retreats as far into the lorica as possible. When recovered from its fright it slowly ascends to the opening, expands itself and resumes its fishing operations. It is fastened to the extreme end of the lorica by the tip of the body; from the sides it is entirely free. On its front border it has a wreath of fine cilia in continuous motion when the animal is extended. The body is

Fig. 111.—Vaginicola.

soft and flexible, and is sometimes of a pale greenish tint, but the lorica, I think, seldom changes color with age. It is not uncommon to find two bodies in one sheath, where they seem to live together in peace and harmony. This may be an advantage to both, for two wreaths of cilia can, of course, produce stronger currents, and so bring more food to the mouths of the always hungry creatures. *Vaginicola* is quite common on Lemna and Myriophyllum.

7. PLATÝCOLA (Fig. 112).

The lorica is flattened, and is in outline almost circular. It is always adherent to some submerged object by the broad flat side, the opposite or upper surface being convex. The opening, through which the animal extends itself as in Vaginicola, is at one end, and is often prolonged into a short neck. The figure shows a side view with the animal extended. When young the lorica is colorless, but it very soon changes to a deep brown, often becoming so opaque that the body of the Infusorium cannot be seen through its walls. The body is usually colorless; it is attached by its tip to the side opposite the mouth of the lorica. When frightened it darts back into the shell as Vaginicola does. Two animals are not seldom found in one lorica. It is not uncommon on Ceratophyllum and other aquatic plants.

Fig. 112.—Platýcola.

8. COTHÚRNIA (Fig. 113).

The beginner may mistake this for a small Vaginicola, as the loricæ somewhat resemble each other in shape; but *Cothúrnia* can always be known by the little stem or foot-stalk that lifts it a short distance from the plant to which it is attached. This foot-stalk in some species is very short, and must be especially looked for. The lorica is vase - shaped, often with the sides variously curved. It changes to a brown color as it grows old. The body of the enclosed Infusorium·is not colored. In its actions it resembles Vaginicola and Platycola, being similarly attached to the posterior end of the lorica, and having a similar circle or wreath of cilia around the front border. Two animals are sometimes found in one lorica.

Fig. 113. Cothúrnia.

9. STÉNTOR (Figs. 114, 115, 116).

The *Sténtors* vary a good deal in shape in the same

species, the bodies of all being somewhat changeable in form. The largest ones are trumpet-shaped, and are usually attached to some object by the narrow end of the body. They also commonly form a soft, brownish, granular sheath or lorica, to the bottom of which they retreat when disturbed, folding together the wide trumpet-shaped front border. The entire surface of the body in all the species is ciliated, but the cilia are very small and fine. Around the edge of the front border is a circle of longer and larger vibratile hairs, visible with a moderately low power. The Stentors are all common. The following Key may help the beginner to recognize some of those most frequently seen.

Key to some species of Sténtor.

1. Attached, and usually forming a short, soft sheath (*a*).
2. Free-swimming, more or less ovoid; green, red, blue or almost black (*b*).

Fig. 114.—Sténtor polymórphus.

  a. Body large, trumpet-shaped, greenish; often without a visible sheath, and when one is formed it is sometimes soon abandoned, the Stentor swimming about freely. The body is slightly changeable in shape. Several Stentors of this species are often found close together, having formed a very soft sheath divided into irregular compartments, one for each Infusorium. *S. polymórphus*, Fig. 114.

a. Body long, and narrowly trumpet-shaped, the front
divided into two lobes, one of which is almost at
right angles to the other.   The body has many
long, fine hairs projecting from it, and
visible under a high-power ($\frac{1}{4}$ inch) objective.   The sheath is always present.
It is narrow, cylindrical, brown, and
about one-half as long as the extended
body.   This Stentor is never free-swimming, and is never found in company
with others of the same species.   It is
not uncommon on Ceratophyllum.   *S.*
Barrétti, Fig. 115.

Fig. 115.—
Sténtor
Barrétti.

b. Body green or red, the red color often being lim-
ited to the part just beneath the wide front border where the circle of large cilia is.   Sometimes
the red color is diffused over the whole body, but
usually the green matter so obscures it that it is
invisible.   This species is often extremely abundant at the bottom of
shallow ponds in early spring.   The
green color then always entirely conceals the red.   *S. igneus*, Fig. 116.

Fig. 116.—
Sténtor
igneus.

b. Body large, indigo blue.   This in shape resembles
Fig. 114 when extended; when contracted it is
not unlike Fig. 116.   Very common in some localities.   *S. cœrúleus.*

b. Body dark brown, almost black.   This also resembles Fig. 116.   Common.   *S. niger.*

10. Astásia (Fig. 117).

Body long and narrow, very soft, and changeable in shape, altering its form as it glides over the slide, which it does quite rapidly. It has one long straight flagellum at the front. It is quite common.

Fig. 117.—Astásia.

Fig. 118.—Eugléna.

11. Eugléna (Fig. 118).

Body long and rather narrow, being widest in the middle and tapering to both ends. It is very changeable in form, and bright green or red in color. The front end is seen with a high power to be notched as if the Infusorium had two lips, the long, vibrating, and colorless flagellum appearing to issue from the notch. There is sometimes a small red spot near the front end, supposed to be an imperfect eye. It is often absent in an old *Eugléna*. At the posterior end is a short, pointed, stiff, and sometimes curved tail, which is usually colorless. The Infusorium is common, occasionally occurring in such immense numbers that it tinges the water green. There is another species, or another variety of this species, whose body is bright crimson. It also is so abundant at times that it colors the water blood red.

12. Chilómonas (Fig. 119).

This colorless little creature is very common in vegetable infusions. It may be recognized by the notch at the widest or front end, and the curve of the back which makes it look almost hunch-backed. Under a high power it shows two flagella, one of them throwing itself into a coil or loop when the Infusorium settles down to rest, which, by-the-way, it quite frequently does. The body is filled with small colorless disks which the iodine solution turns blue, showing that they are starchy.

Fig. 119.—Chiló-
monas.

Fig. 120.—Phácus
pleuronéctes.

Fig. 121.—Phácus
longicaúdus.

13. Phácus (Figs. 120, 121).

The body of *Phácus* is flattened, thin, and rather like a small leaf. It is widest in front, usually rounded, and tapering from the centre to the short, pointed, colorless tail-like prolongation; at the broad end it has one long flagellum, often difficult to see. There are several species in our ponds, all of which are green.

1. Body not twisted at the rear, tail short, curved. *Ph. pleuronéctes*, Fig. 120.
2. Body twisted or not at the rear, tail long, straight. *Ph. longicaúdus*, Fig. 121.

14. Uvélla (Fig. 122).

The little animals forming these rapidly swimming and revolving colonies are united by their narrow ends into almost spherical microscopic masses, varying in

number from two or three up to forty or fifty or more.
Each Infusorium has a narrow, yellowish - green band
down each side of the somewhat egg-shaped body, and
two long, fine flagella at the broader front end. The
colonies are common in early spring in shallow pools
with Algæ.

Fig. 122.—Uvélla.

Fig. 123.—Trachelocérca.

15. TRACHELOCÉRCA (Fig. 123).

This will probably be a greater surprise to the beginner the first time he sees it than any other common Infusorium, on account of the remarkable neck, which can
be stretched out to five or six times the length of the
body, and drawn back until it almost entirely disappears. The body, without the neck, is somewhat spindle-shaped, and occasionally ends in a short, tail-like part.
The Infusorium may often be concealed in a mass of
fragments or a heap of dirt, while only that wonderful
neck is visible, stretching and bending and writhing like
a colorless snake, as it searches the slide for food. The
end of the neck is rather pointed, and bears the mouth
at the tip. The whole Infusorium is covered with fine
cilia. It is quite common.

16. AMPHILÉPTUS (Fig. 124).

This is one of the largest of the Infusoria, sometimes

measuring $\frac{1}{16}$ inch in length. The neck is not extensile as in Trachelocerca, although it is the longest part of the

Fig. 124.—Amphiléptus.

whole animal. The body, without the neck, is somewhat spindle-shaped, tapering more rapidly towards the rear than towards the front. The latter or neck-like part is flexible, and is turned and twisted about in a way that often suggests the movements of an elephant's trunk. The whole body is covered with fine cilia.

17. PARAMÆCIUM (Fig. 125).

This is often called the "slipper animalcule" from its shape. It is frequently found in the ponds, but is especially abundant in vegetable infusions. The hollow place resembling the opening in the slipper for the foot, is the part leading to the mouth near the centre of the lower surface. The whole body is cov-

Fig. 125.—Paramæcium.

ered with fine cilia, and sometimes a cluster of longer, coarser cilia is noticeable on the posterior tip of the body. In the writer's locality this cluster of cilia is present on all the specimens; I have never seen a *Paramœcium* without it. This Infusorium increases rapidly by dividing into two parts across the middle. Its movements are rapid.

18. EUPLÓTES (Fig. 126).

This is one of the walking Infusoria, the cilia on the flat lower surface being very large and strong. The animal uses them for swimming, or it walks about the slide or climbs among aquatic plants by resting part of its weight on their tips as if they were legs. When the creature happens to be turned on its back, these large cilia can be seen pattering irregularly against the cover-glass. They vary in number from ten to twelve. The front border has a row of finer but still large cilia extending down the side of the flat surface to the mouth near the centre of the body. Four straight, stiff hairs project from the posterior margin, two of them often being divided into fine branches. The back of the Infusorium has no cilia, but is a hard surface, almost like a shell. The animal is very active. There are several species common among Ceratophyllum and Myriophyllum.

Fig. 126.—Euplótes.

19. STYLONÝCHIA (Fig. 127).

To the beginner the members of this genus will quite closely resemble Euplotes, as all the cilia are confined

Fig. 127.—Stylonýchia.

to the frontal border, the part about the mouth, and irregularly distributed over one side of the flat lower surface as walking organs. It can easily be distinguished from Euplotes by its shape, being much more oblong. Sometimes it is quite long and narrow, while Euplotes is always more or less circular. It has no cilia on the back, which is usually hard and shell-like. The species are several, being especially common in vegetable infusions.

20. Chílodon (Fig. 128).

The body is oval and flattened, the lower or flat surface alone being ciliated. The front border is convex,
and rather sharply pointed at one corner, and the side
of the body extending from this corner to the rounded
posterior margin is nearly straight, while the opposite
side is convex. The back is smooth and naked.
From the pointed corner a curved line of cilia
extends back over the flat surface to the mouth,
which opens into a cone-shaped bundle of fine
rods visible under a high power. The ends of
these rods can be seen with a moderately low power, en-
circling the mouth like beads. The Infusorium lives
upon smaller Infusoria and diatoms, which it appears to
seize with this peculiar throat, the rods separating as the
food is slowly swallowed. *Chilodon* is common in still
waters.

Fig. 128.
Chilodon.

21. Loxódes (Fig. 129).

The body is quite long and narrow, the frontal border
being convex, with one corner rather pointed; but on one
side, just below the pointed corner, is a concave space containing a brown, sickle-shaped body lining the
hollow which is part of the Infusorium's throat.
The upper portion, or blade of the sickle, seems
only to stiffen that part of the cavity, the true
mouth being at the beginning of the short handle
of the sickle. The cilia are fine, and are on the lower flat surface only. The body is flexible, often bending on
itself. The Infusorium is quite common in some localities.

Fig. 129.
Loxódes.

CHAPTER VI.

HÝDRAS.

WHEN Hercules was going about doing those wonderful things of which we have all heard, it was suggested that he should turn his attention in the direction of Lake Lerna, near Argos, where a monster with a hundred heads was making itself unpleasantly active. He visited the place and interviewed the creature, but when he had cut off one of the heads, he must have been surprised to see two new ones sprout out of the bleeding surface. It was discouraging, but the hero began to have the best of the contest when he began to burn the fresh cuts with a hot iron. The monster was the Hydra of mythology. Science has preserved its memory by giving the name to a common and curious creature inhabiting all our ponds and ditches. The fresh-water *Hydra* (there are no salt-water Hydras) has a soft and elastic body attached by the tip of one end to an aquatic plant or other submerged object, and eight or ten long fine arms arranged around a mouth at the opposite end.

There are two species, the green (*H. víridis*) and the brown (*H. fúsca*), both being very common. The whole animal is elastic, and when extended may be an inch long and easily visible to the naked eye; when con-

8

tracted it resembles a minute globule of green or brown
jelly, with the shortened arms at the summit like very
small drops or projections. It is very active so far as
the arms are concerned, for the body is always adherent
to some submerged object. The arms or tentacles are
usually stretched out to their fullest extent, then often
exceeding the body in length, waving and twisting about
in search of prey. The figure (Fig. 130) shows several

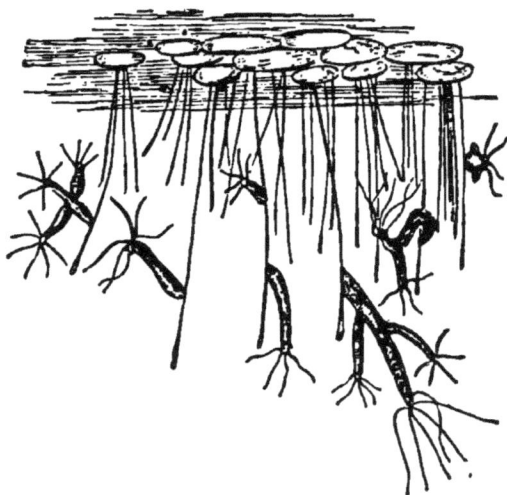

Fig. 130.—Hydras adherent to Lemna rootlets.

Hydras nearly the natural size adherent to Lemna rootlets. The body is like a narrow bag, the hollow part
of the little sack being the stomach, and communicating
directly with the external water, in which the Hydra
lives, by means of the mouth, around which are arranged
the arms or tentacles. These tentacles are themselves
hollow, and communicate with the hollow of the stomach.

The food consists of small worms, water-fleas, or other
Entomostraca (Chapter X.), or even little pieces of raw
beef, if the observer chooses to feed them. They seize
the victim as it is swimming past, by twining a tentacle
around it and drawing the struggling creature down to
the mouth, through which it is thrust into the stomach.
The act of seizure takes place so rapidly that the eye
can seldom follow it. The observer can usually only
know that the prey is caught and is slowly approach-
ing the mouth. Often when the captured object is too
large or strong for one arm to hold, several tentacles
bend over and twine around it. A creature once caught
rarely escapes. When a quantity of aquatic plants is
brought home, the Hydras soon make their way to the
lightest side of the aquarium or bottle and attach themselves to the glass. At such times I have often amused
myself, and doubtless pleased the Hydras, by feeding
them with small larvæ or aquatic worms. Take with
the forceps a small aquatic worm by one end, and present the wriggling thing to a Hydra's arm. No second
invitation is needed. The worm is embraced as quick
as a flash, and, if too long to be swallowed all at once,
part of it will hang out of the mouth until the other
end is partially digested, but the tentacles will not cease
to fish for more. It is said that if the Hydra and the
worm are placed in a very deep cell under the microscope, the performance can be watched through a low-
power objective. I have never succeeded in doing this,
but there is no trouble in feeding them in an aquarium.

They never eat any but animal food, and they are always hungry.

The body and tentacles of *Hydra viridis* are roughened by little elevations or warty prominences. The brown species (*H. fusca*) is not so much roughened. These warts contain what are called the stings. These are small oval or vase-shaped hollow bodies, with a fine thread coiled in the interior, and four minute spines near the summit. When the Hydra is irritated by the pressure of the cover-glass these stings are thrown out violently, and the long stiff thread can be well seen. When in the animal's body they cannot be easily examined. One is shown much magnified in Fig. 130*a*. They are often found on the slide when no Hydra is to be seen, and they are sometimes noticeable sticking in the body of some worm or larva that has escaped a fatal embrace. I have more than once found a Chirónomus larva (Chapter VII.) in a dying condition and ornamented by a spiral band of these stings in its skin, it having evidently had a tussle with a Hydra and escaped.

Fig. 130*a.*
Hydra sting.

. The Hydra increases in numbers rapidly by a process of budding. A little protuberance appears on one side of the body, enlarging and growing, and finally, while still attached to the parent, developing tentacles, then resembling the mature animal in everything except size. And it is not unusual to see one or more still younger Hydras sprouting from these before they are free from

the parent, so that the old Hydra is often a grandmoth‍-
er before she is a mother. The young one is hollow,
and communicates with the hollow of the parent. It
captures food like the parent, and it is said to be no uncommon sight to see the old and the young both seize
the same worm. In such cases the strongest wins, unless the worm breaks in the unfilial struggle, when the
parts go into the one common stomach. Very often
two young Hydras may be noticed growing from the
sides of a single older one, instances of which are shown
in Fig. 130. The budded young finally separate from
the parent, then leading an independent life, and soon
producing young Hydras from their own sides, if they
have not already done so.

The creatures are very hardy. They may endure
much harsh treatment, and seem to thrive under it.
They have been made the victims of many apparently
cruel experiments, but they are probably not very sensitive to a feeling of pain. The sensation of hunger,
and a touch delicate enough to know when a desirable
morsel or an obnoxious object comes in contact with the
tentacles, are probably the extent of their feelings.
Trembley, a Dutch naturalist who studied the Hydra
as long ago as 1739, first called attention to the harsh
treatment they would endure and live. In a rather
quaint, old-fashioned translation it is said that, "If one
of them be cut in two, the fore part, which contains the
head and mouth and arms, lengthens itself, creeps, and
eats on the same day. The tail part forms a head and

mouth at the wounded end, and shoots forth arms more
or less speedily as the heat is favorable. If the polype
be cut the long way through the head, stomach, and
body, each part is half a pipe, with half a head, half a
mouth, and some of the arms at one of its ends. The
edges of these half pipes gradually round themselves
and unite, beginning at the tail end; the half mouth
and half stomach of each becomes complete. A polype
has been cut lengthwise at seven in the morning, and in
eight hours afterwards each part has devoured a worm
as long as itself." He also sliced them across, and found
that each piece developed a cluster of tentacles, and he
finally turned them inside out, and in a few days the
maltreated creature swallowed food, although its old skin
was now lining its stomach, and its old stomach membrane had now become its skin.

There is a peculiar parasitic Infusorium (Fig. 130b)
often seen in considerable numbers gliding rapidly over
the body and arms of the Hydra, especially of *H. viridis*. They do not seem to be objectionable
guests, as the Hydra never appears to notice
them. It is said that they infest sick or
weakly victims only, but that is not according
to the writer's experience, if the condition of
the Hydra may be judged by appearance, ac-
tivity, and appetite. One of these parasites
is shown in side view (Fig. 130b). It is shaped like
a short dice-box, with a circle of fine cilia at each
end, but none on the rest of the body. It glides along

Fig. 130b.
Trichodina
pediculus—
Parasite of
Hydra.

rapidly on the ends of the dice-box, running out to the tips of the tentacles and skirting fearlessly around the edges of the mouth. It is the *Trichodina pediculus*.

The Hydra also occasionally has another form of Infusorial parasite running over its skin. This is somewhat kidney-shaped, and has cilia only on one surface of the body. It is called *Keróna polypórum*. It does not seem so common as Trichodina.

If the observer desires to preserve the Hydra as a permanently mounted object for the microscope, he may be easily gratified, thanks to Mr. A. II. Breckenfeld,* of San Francisco, who has devised an admirable method which the writer has tried and recommends. Transfer the Hydras to a slip in a large drop of water, where they can be seen if the slide is held over white paper. When their tentacles are fully extended, "quickly move the lamp directly under the drop, with the top of the chimney about an inch beneath the slide, and hold it in that position for about three to five seconds, the exact time depending principally upon the intensity of the heat. Then quickly remove the slide and place it upon a slab of marble or metal. When cool, pour the drop containing the zoophytes into the prepared cell on the slide which has been held in readiness; add a drop or two of a suitable preservative fluid, arrange the little animals if necessary by means of a needle or camel's-hair brush (using very great care, however, as the ten-

---

* *American Monthly Microscopical Journal*, March, 1884, p. 49.

tacles will be destroyed by the least rough handling), cover with thin glass, and finish as in the case of any fluid mount." I have not found it necessary to use two slips of glass. If a deep shellac cell that has been made for some time and is perfectly dry and hard is used, the Hydras may be placed in it and there cooked and allowed to remain. When cold, arrange the arms if necessary, add a drop of weak glycerine and water, and cement the cover-glass with shellac. The Hydras thus prepared can be kept indefinitely, and at any time shown to admiring friends.

Both the green and the brown species are abundant during the summer among Anacharis and Lemna.

CHAPTER VII.

SOME AQUATIC WORMS, CHÆTONÓTUS, AND CHIRÓNOMUS LARVA.

THE collector of microscopical objects from the ponds and slow streams is doubtless familiar with the appearance of the bristle-bearing worms (Fig. 140), on account of their general resemblance to those long-suffering creatures which he in his youth impaled on a hook and with them sought the nearest water. The extensive bristles of the aquatic worms are an addition which greatly lessen their resemblance to the common earthworm, and their transparency is another characteristic that may temporarily mislead the observer, but their elongated bodies and general worm-like aspect tell the story. In addition to the bristles which most members of this class possess, there are usually two or more rows of long, curved spines (Fig. 141) on the ventral or lower surface. These can be protruded or withdrawn into the body at the possessor's will, and when protruded are used to assist the worm to crawl. They are therefore called the podal or foot spines. They may not be noticed when retracted unless specially searched for. Having observed them and the bristles in a row on each side above them, the student need have no trouble in knowing where to class the worms; but with an-

8*

other division of the group the beginner may not fare so well.

These have flattened, usually almost opaque bodies, with the entire surface densely clothed by fine cilia, and, probably on account of the stir and disturbance which the cilia make in the water, naturalists have classed the worms together under the name of the *Turbellária*, from a Latin word meaning a stir or bustle. Their motions are rapid, and apparently without effort. They glide smoothly and swiftly over submerged objects, or swim back downward on the surface of the water.

There is still another group of common aquatic worms, but to recognize them will give even the beginner very little trouble. They are often rather sluggish in their movements. They have a perfectly transparent, smooth, thread-like body, which is apparently truncate in front, and is prolonged posteriorly in a sharpened, point-like tail. They have no bristles nor cilia, and they rather closely resemble a microscopic eel; indeed the scientific name, *Anguillula* means a little eel.

Many members of all these classes are found in the superficial sediment of shallow ponds, in the crevices of wet and water-soaked logs, under submerged stones, among the leaflets of Myriophyllum, Sphagnum, and other water-plants. Sphagnum seems a favorite place for several kinds. I have obtained members of five genera, *Náis, Pristina, Déro, Chœtogáster,* and *Æolosóma*, by placing a little piece of the moss in a watch-glass with a small quantity of water, and gently tearing away the leaves with needles, when the concealed worms hurried out and were readily captured with the dipping-tube. If the watch-crystal stands on black paper the work is facilitated, as the translucent worms then appear to the naked eye as minute, writhing, silvery threads.

In this chapter the reader will also find descriptions of two very common microscopic aquatic animals, one of which is certainly not a worm, the proper position of the other being rather doubtful. They are *Chœtonótus* and *Chirónomus* larva (Figs. 131, 132), both having somewhat worm-like bodies. They are here referred to for the convenience of both reader and writer. The beginner will be quite sure to at first mistake Chironomus larva for a worm.

The bodies of all the worms are very soft and easily injured. It is best, therefore, in studying them to use a cell shallow enough to somewhat restrain their movements, when the cover-glass is added, but deep enough to avoid undue pressure, or they will rapidly go to pieces.

The following Key will assist the beginner in determining to which class his worm may belong, leading to the names of the groups under which some of their generic titles may be found:

1. Body with four leg-like appendages bearing hooked bristles; eyes distinct; head large, brownish-red. *Chirónomus larva*, I.

2. Body without leg-like appendages (*a*).

   a. Tail forked; mouth small, circular, on the front

part of the lower or ventral flat surface; back convex, usually bearing spines, prickles, or scales. *Chætonótus*, II.

a. Tail not forked, but often bearing finger-like appendages (*b*).

b. Body entirely and finely ciliated, usually flattened. *Turbellúria*, III.

b. Body smooth, without cilia, bristles, or spines; tail pointed. *Anguíllula*, IV.

b. Body with bristles, podal spines, or both. *Oligochæta*, V.

I. CHIRÓNOMUS LARVA (Fig. 131).

Chirónomus larva has a worm-like, more or less jointed, colorless body, eight or nine times as long as wide, a large head, the mouth parts usually being distinctly apparent. The four short rudimentary leg-like appendages are in pairs on each end of the long body, the brownish hooks or strong curved bristles on their extremities being more or less retractile, while two clusters of long bristles spring from the upper surface near the posterior border of the animal. The perfect insect into which this larva will develop is a two-winged fly resembling a mosquito. These are often seen in great numbers above the ponds and marshes. The species are very numerous, and have never been studied by American entomologists.

The eggs are very common on sticks, floating chips, or other objects in the water. They are deposited in a

mass of jelly, huge in bulk when compared with the
size of the insect, the eggs appearing as distinct but
minute, often brownish,
specks, arranged in beautifully regular rows.

It is always interesting
as well as important for
the collector to take home
all the little jelly-like egg

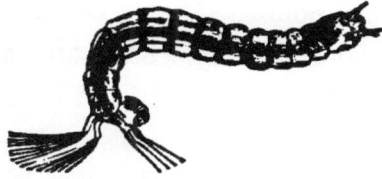

Fig. 131.—Chirónomus larva.

masses which he may find attached to submerged objects.
If placed in a watch-glass or an "individual" butter-dish,
and the water kept fresh and pure, they will usually
hatch, and thus give the observer valuable information
often not otherwise obtainable.   Chironomus eggs can
hardly be described so that the beginner shall recognize
them at first glance, but if once hatched at home they
will afterwards always be known.   The first little mass
of jelly experimented with may prove to be snails' eggs,
but they will be none the less interesting.   They may
also prove to be the eggs of water-mites (Chapter XI.).
The beginner will, of course, not mistake the green jelly
globules of Chætophora for insect eggs.

II. CHÆTONÓTUS (Fig. 132).

There are several species of these lithe and graceful
little creatures in our fresh waters, and they so closely
resemble each other in external form that they can
be distinguished only by the cuticular appendages, or
the coat-of-mail by which most of them are protected.

They are readily to be found by fishing for them with a dipper, as recommended for Rhizopods, as they are fond of gliding over the soft ooze at the bottom of shallow ponds. If the collector will also sweep his dipper under the lily leaves and among the submerged stems of Nuphar, he will not be disappointed.

The animal consists of a free-swimming, flexible, and elongated body, the anterior extremity usually enlarged to form what may be called the head, a slight constriction behind this part constituting the neck; the central portion of the body is formed with convex lateral borders and a more or less strongly convex back or dorsum, this region being variously appendaged with spines or scales, and suddenly narrowed to produce the posterior extremity, which is forked, and bears two conspicuous tail-like prolongations. The lower or ventral surface is a flat and nearly level plane extending the entire length of the body. It bears one longitudinal band of cilia near each lateral border, seldom more. The head is usually somewhat triangular, and formed of three or five rounded lobes. It has two tufts of vibratile hairs on each side.

Fig. 132.—Chætonótus larus.

The mouth is on the ventral surface of the head, and under a moderate amplification seems to be a circular opening, but with an objective of high power it will be found to be somewhat complicated.

The whole upper surface of the body is, in the different species, covered with rounded papillæ, scales, spines,

or prickles, or with both scales and spines at the same
time.　In the latter kinds the scales cover the back and
sides, and the spines spring from these appendages, arch-
ing back towards the forked tail.　And in all cases these
little scales are imbricated, or overlapping like the shingles on a roof, only they have the curious habit of lap-
ping in what seems to be the wrong way, that is, their
free margins point towards the animal's head, or in a
direction just opposite to that of the scales on a fish.
They are usually minute, and require high powers to
see them properly.

The two caudal prolongations are movable and flexible.　Their chief use seems to be to anchor the animal
to the glass slide or cover, or to some object in the water, clinging with the tips, and apparently assisted by
a secretion that is supposed to exude from them, this
sticky fluid passing from two ovate glands usually visible in the upper or anterior part of each.

The mouth opens into a very muscular œsophagus,
which itself opens into the intestine, a tapering, tubular
passage lined with nucleated cells and passing in almost
a straight course along the median line, terminating between the two caudal prolongations.　If the observer
can get the animal in such a position that he can focus
down on the front of the head, he will see that the cavity of the œsophagus is triangular.　It is not very difficult to do this, since the little creatures are exceeding-
ly restless; they are continually turning and writhing
about, and lifting the head in various directions.　It can

often be seen in the animal while still in the egg, for
even there, when almost ready to escape, it is also very
restless.   The eggs are often found on the slide, with
the young Chætonotus doubled up within.

The eggs by which Chætonotus is reproduced are
formed in an ovary placed in the median line of the
body immediately above the intestine.   Usually only
one egg is formed at a time, but it is not rare to see two
or more in various stages of ovarian development.   Upon
the absence or presence of an egg in the ovary depends,
to a great extent, the degree of convexity of the back.
The eggs are dropped anywhere in the water, and left
to the care of nature.

The food consists of the minute particles of decayed
animal and vegetable matters so abundant in the soft
surface of the mud at the bottom of our shallow ponds.
These particles are taken in with a peculiar and a sudden snapping movement of the cavity of the œsophagus,
easily to be seen but difficult to describe.   Diatoms are
rarely swallowed.

So far as their classification is concerned, these attractive little animals have given naturalists a good deal of
trouble.   Some have said that they belong with the
Rotifers; others have placed them among the Infusoria; others have called them low worms, putting them
among the Turbellaria; and still others think, and they
are doubtless correct, that Chætonotus should stand in a
group by itself, among the worms, and not very far from
the Rotifers.

They are all rapid swimmers, and on that account are rather difficult to study, but by following one for a little while, it will usually settle down and begin to seek food, and that is the observer's opportunity, unless he desires to kill the specimen, and study it after death.

The following Key leads to some of our common forms:

Key to Species of Chætonótus.

1. Upper surface with neither spines, prickles, nor scales (*a*).
2. Upper surface bearing scales only (*b*).
3. Upper surface bearing spines or prickles only (*c*).
4. Upper surface bearing both spines and scales (*f*).
5. Upper surface bearing posterior spines and anterior prickles (*g*).

    a. Back smooth and naked, not furrowed, *podúra*, 1.

    a. Back transversely furrowed, *sulcátus*, 2.

    a. Back covered with small hemispherical elevations, *concínnus*, 3.

    b. Caudal branches of moderate length, scales rounded, *loricátus*, 4.

    b. Caudal branches very long and jointed; scales very small, rhombic, *rhomboídes*, 5.

    c. Spines covering the entire upper surface (*d*).

    c. Spines not covering the entire upper surface (*e*).

    d. Spines long, mouth beaded, *máximus*, 6.

    d. Spines short, mouth not beaded, *lárus*, 7.

    e. Spines eight, in two longitudinal rows of three

each, with one anterior and one posterior central spine, *octonárius*, 8.

e. Spines in two transverse rows, not projecting beyond the ends of the caudal branches, *spinósulus*, 9.

e. Spines in two transverse, highly-arching rows, the posterior longest and projecting beyond the ends of the caudal branches, *longispinósus*, 10.

f. Back with a subcentral, transverse hedge of large spines, scales double, *acanthódes*, 11.

f. Back without a distinct spinous hedge, scales not double, *spínifer*, 12.

g. Spines in four transverse rows, five spines in each, *acanthóphorus*, 13.

g. Spines in transverse rows, less than five spines in each, *enórmis*, 14.

1. CHÆTONÓTUS PODÚRA.

Chætonótus, or bristle-back, is rather a misnomer for a species with a perfectly smooth dorsum, yet such a one is not uncommon. The spines and other dorsal appendages are here represented by two hairs standing almost vertically on the neck, and two on the rear part of the back. These are usually seen with difficulty, but they are present on all the species, even the scaly and the spinous ones. The egg of this species is also smooth. Ehrenberg called this *Icthýdium podura*.

2. Chætonótus sulcátus.

The characteristic of this form is in the deep transverse furrows conspicuously developed on the back and sides. The body is transparent, and unusually soft and flexible. The posterior region between the arch of the back and the caudal furcation is narrowed, and much longer than in other species. The œsophagus is short, being not more than one-sixth the length of the body.

3. Chætonótus concínnus.

The back and sides, which are more nearly parallel than in most species, are closely covered by small hemispherical elevations arranged in oblique lines and giving the animal a peculiarly neat and attractive appearance. The two caudal glands are unusually large and conspicuous.

4. Chætonótus loricátus.

The scales on the back and sides are arranged in imbricated rows, the convex free margins being directed forward. Although so completely covered, the body is very flexible, the scales freely sliding over each other when the animal curves to one side. The mouth is obliquely placed, as may be seen when the Chætonotus is viewed in profile, and its internal margin is strongly beaded. The eggs are armed by hollow papillæ, or by short hollow spines whose summits are bifid or emarginate.

5. Chætonótus rhomboídes.

This is easily recognizable by the peculiar head, the

minute rhombic scales covering the back and sides, and by the remarkably long and jointed caudal branches, each of the latter forming from one-third to one-fourth of the entire length of the body. The animal is the largest yet discovered, measuring $\frac{1}{85}$ inch long. The caudal branches are composed of about twenty sections or joints, each of which is slightly constricted. The head is broadly rounded, and formed of three lobes, one frontal and two lateral, the former terminating on each side in a single, acuminate, hook-like process, habitually in close apposition with the anterior region of the lateral lobes, of which the posterior extremities also terminate each in a single hook-like continuation, rather more conspicuous than those at the front. The mouth is beaded, and has immediately behind it on the ventral surface a deep, narrow, transverse, and slit-like depression, rather less than one-half as long as the diameter of that part of the head. This is the only known Chætonotus with this problematical feature.

The back and sides are completely clothed by minute, imbricated, rhombic scales, their front, pointed margins being directed towards the head. They are not more than $\frac{1}{5000}$ inch in length, and when examined with a high power (one thousand diameters) they present a beautiful appearance. The lateral margins then seem to be thickened, and the posterior border of each scale appears to bear a minute supplementary scale in the shape of a triangle.

Although the beginner may not be able to distinctly

see these scales, the very long caudal branches with their joints, and the sulcation behind the mouth, will be sufficient to identify the specimen.

6. Chætónotus máximus.

The back and sides are covered with spines which are often rather longer on the posterior region than elsewhere. They are arranged in longitudinal parallel rows, yet they often seem to be irregularly scattered, so that the animal presents an untidy, dishevelled, and disreputable appearance. The spines are minutely forked near the free ends. The branching is very uneven and is easily overlooked, one branch being very small, often scarcely more than a minute linear projection.

The ventral cilia are in two longitudinal lateral bands, and the space between is clothed with short, hispid, recurved hairs, two or more long fine bristles projecting from the same part beyond the posterior border, between the two caudal branches.

7. Chætonótus lárus (Fig. 132).

The whole upper surface is clothed with short, conical spines in longitudinal rows, these appendages being recurved and not branched. They are often largest posteriorly. The mouth is not beaded. The ventral cilia are in two broad longitudinal bands near the lateral margins, and the intervening space often bears two additional parallel lines of cilia, which may be absent from some specimens. These cilia, as in all the species, subserve locomotion.   The egg is smooth, or hispid with short hairs.

8. Chætonótus octonárius.

This is a small, active form, readily recognizable by the arrangement of the recurved dorsal spines.   These are unequally branched, and placed in two lateral longitudinal rows of three spines each, with one anterior and one posterior central thorn.   It seems to be the least common of the species.

9. Chætonótus spinósulus.

The back usually bears seven unequally furcate spines in two transverse rows—four spines in the anterior series, three in the posterior.   Occasionally the lateral thorns in the posterior row are suppressed, and in some individuals the front series contains but three.   The lateral body-margins are bordered by short, conical setæ, which are constant in all the specimens thus far observed.   The rest of the upper surface is without appendages of any kind, except the four tactile vertical bristles present in all species.   The egg is hispid with short hairs.

10. Chætonótus longispinósus.

The spines vary from four to eight, the latter being the usual complement.   They are nearly one-half the length of the body, and curve upward and backward in a wide arch from the centre of the back.   In front of the anterior row the surface is setose with stiff, recurved bristles, and the body-margins are fringed by coarse, rigid setæ. The dorsal spines are always in two transverse rows, but the number varies from four in each to three in one and five in the other. They are unequally furcate.

11. CHÆTONÓTUS ACANTHÓDES.

The upper surface of this form is wondrously well protected. It possesses both spines and scales, the latter imbricated, and their somewhat pointed free margins directed forward, each one bearing a small supplementary scale or scale-like thickening on its posterior part, from which springs a recurved, unequally furcate spine. Near the body-centre the dorsal surface is traversed by a series of large stout spines rising obliquely upward and backward, and forming a kind of spinous hedge, the surface behind these appendages bearing few small conical thorns or none. The body margins are fringed by short spines. The central space on the ventral aspect between the two longitudinal, lateral bands of cilia, is beset with short, fine, recurved prickles, and five or more long bristles project from the same surface beyond the border of the posterior bifurcation, while on each side of the body near the posterior extremity there are two large recurved spines. The animal is usually found among Sphagnum.

12. CHÆTONÓTUS SPÍNIFER.

Among Riccia and Lemna in shallow ponds this well-armored form is not rare. The upper surface is covered by rounded imbricated scales, the free margins directed forward. From each scale there arises a stout, recurved, unequally and minutely furcate spine, whose base is enlarged and thickened. These spines do not commonly originate from the centre of the scales, but near the posterior part, and between the margins of those laterally contiguous. The spines are largest and stoutest on the back proper, decreasing gradually over the neck and head, and rapidly over the posterior parts, while across the dorsal surface immediately in front of the caudal bifurcation there extends a supplementary series of four thorns, longer and stouter than those on any other part of the body. The posterior region of the space between the longitudinal ventral bands of cilia bears five bristles, arranged to form a long triangle, the apex pointing forward.

The eggs vary considerably in external ornamentation, showing three patterns. In one, the ends and one side bear low, stout, hollow processes, whose apices are truncate, and four or five parted when viewed from above. In another, the appendages are long, hollow, conical spines, whose distal ends are trifid or quadrifid, the branches in profile appearing very fine and delicate, but when viewed from above are seen to taper to the ends, where each terminates in a widely spreading furcation. In the third form, one side and both ends are covered by an irregular net-work of raised lines, the meshes being four or five angled, while the opposite side is rugose with fine, minutely sinuous lines.

13. CHÆTONÓTUS ACANTHÓPHORUS.

The superior surface of the head and neck and the lateral body-margins are clothed with recurved prickles or short spines, while the dorsal region proper bears four rows of long thorns, each row curved towards the head, and each formed of five unequally furcate spines, with an additional one on both sides near the posterior extremity. The spines rise from an enlarged base, so that the animal is almost completely clothed in an armor composed of these basal enlargements.

14. CHÆTONÓTUS ENÓRMIS.

The upper and lateral surfaces of the head and neck are clothed with short, recurved prickles, which also extend along the ventro-lateral margins. The central and posterior parts of the back bear thirteen posteriorly directed, but only slightly curved, spines arranged in transverse rows, with three in the first row, four in the next following, two widely separated in the third, three in the fourth, while the fifth series consists of a single centrally located one. On each side near the posterior margin are two long, conspicuous, and recurved thorns, apparently belonging to the series of small spines fringing the lateral body-margins.

III. TURBELLÁRIA.

The ciliated or *Turbellárian* worms seem to prefer the bottom of shallow ponds, probably because the food-supply there is better and more easily obtained. They

9

are soft and flexible, and some are quite changeable in shape, having the power to lengthen themselves, to extend the posterior border into a short projection, or to narrow the front into an apology for a head. Some, however, have the front part naturally prolonged into a short snout. They are usually brownish and almost opaque, the opacity being increased by the large amount of food commonly present in the stomach.

The cilia clothing the entire surface are visible only under a high power. The result of their motion, however, can be seen with the one-inch objective, as they produce currents in the water that sweep small objects quite rapidly away.

Two or more small black or reddish eye-spots are often present near the front border, and in some of these worms may be quite complicated in structure, having a covering that may not inappropriately be called a cornea, a refracting body corresponding to a crystalline lens, pigmentary or coloring matter, and a nerve.

The position of the mouth varies widely in the different families. It may be at or near the front border, at some point nearer the centre of the body, or even close to the posterior margin. It is usually large and expansile, and is often followed by a large and very muscular organ called the pharynx, which some of the worms can protrude, and with it snap up their living prey. The lining of the pharynx may be finely ciliated.

The stomach occupies the largest portion of the body,

usually extending from the pharynx to the posterior
border. In some it is simply a great sack, receiving all
that the mouth and pharynx turn into it; in others it
divides into many branches whose terminations may be
seen near both sides of the body. The stomach seldom
has a posterior opening, for, as a rule, there is no intestine. After the nutriment has been digested from the
food, the insoluble remains must be ejected through the
mouth. It is no unusual sight to see one of these ciliated worms vomit up a mass of indigestible and empty
Rhizopod shells, Rotifer carapaces, with many unrecog-
nizable particles and fragments. They seem to prefer
animal food, usually selecting Rhizopods and Rotifers,
but they are as fond of Infusoria, which must be as
nourishing and more easily digested. I have more than
once lost an interesting specimen of Infusorium because
one of these Turbellarian worms had been included under the cover-glass: there were a worm and an Infusorium; a pause, a single snap, and only the worm remains.

They are propagated in two ways — by eggs and by
transverse fission; that is, one worm divides across the
middle and so makes two, each of these again dividing.
And often before the division has been accomplished
both halves are also partly divided, so that the single
body seems to be formed of several incomplete worms.
The eggs of the commonest species are brownish egg-
shaped bodies, dropped anywhere in the mud or water,
or they may have a stem which attaches them to sub-

merged objects, and from which they are easily broken. The latter kinds are readily recognized, being formed of a yellowish brown, transparent membrane, egg-shaped, and with the stem almost equalling their own length. If the observer be fortunate he may see the worm escape by pushing off the top of the egg, which falls away like a round cover, leaving an empty case shaped like a deep cup. These empty vases are often found at the bottom of long-standing collections of plants.

The Turbellarian worms are very common, but the beginner can scarcely hope to learn the generic name of each one that he may find. He will be safe, however, if he refers to them all as Turbellarians, or Turbellarian worms. The subject has not been studied very extensively by American naturalists, and there is, consequently, nothing in the language to which the beginner can be referred for help.

The worms are often visible to the naked eye as minute whitish or flesh - colored floating bodies, or as small bits of white thread. There are two forms frequently met with which are huge when compared with most of these ciliated creatures, needing no microscope to identify them. Both are found on the lower surfaces of submerged stones or sticks, or gliding over the sides of the collecting-bottle.

The body of one of these may be about half an inch in length and about five times as long as broad. It is opaque and almost black. Near the anterior border are two black eyes, which are conspicuous on account of the oblong white space in front of each. The mouth is near the centre of the body, opening on the lower or ventral surface. The worm glides smoothly and quite rapidly over a submerged surface. Naturalists have named it *Planária tórva.*

The second one referred to somewhat resembles *Planária tórva*, but is usually smaller, and has the head end more nearly triangular. It is similar in its movements and in the presence of two black eyes near the front border, each at the inner margin of a white space, thus giving the worm a cross-eyed appearance. The body is nearly white, and has a dark line passing lengthwise through the centre and giving off on both sides many short branches which are themselves often branched, these dark lines on the white body giving the latter a very pretty appearance. They are not for ornament, however, but are the branching stomach. The mouth is near the centre of the lower surface. The body may measure half an inch in length. It has been named *Drendocœlum lácteum.*

The entire surface of both these worms is finely and closely ciliated. The color of the body will at once inform the observer which one he has captured.

IV..ANGUÍLLULA (Fig. 133)

The body is thread-like, perfectly transparent and colorless, about fifteen times as long as broad, rather widest in the middle, whence it slightly tapers towards both ends. The frontal border is rounded, but with a low power appears as if truncated. The round mouth is at the centre of this end, and leads into an oblong pharynx or throat. The tail is usually long and sharply pointed. The worm's movements are generally slow and deliberate, but occasionally it has a lively spell. They

Fig. 133.—Anguillula.

are reproduced by eggs, one or more often being visible within the transparent body. Anguillulæ are common in wet moss, among the leaflets of aquatic plants, and in the ooze of the ponds.

The well-known "vinegar-eel" is an Anguillula (*Anguillula acéti*); and the paste-worm (*A. glútinis*) belongs to the same genus. Some naturalists regard these as the same species.

V. OLIGOCHÆTA.

The fresh-water bristle-bearing worms whose bodies are never ciliated, show more or less distinctly that they are formed of segments or rings. Each segment often has on both sides near the back one or more long, fine, hair-like bristles extending into the water, and together forming a row along each side of the worm. On the lower surface there are two or more rows of thicker, inflexible, and gracefully curved spines, the rows being formed of clusters which have two or more spines in each, the free end of every one being usually divided by a deep notch, so that it appears like a double hook, the parts being unequal in size and degree of curvature. They are used to assist the worm to crawl, and are therefore called podal or foot spines. They can be protruded from the body, or partially withdrawn into it, at the animal's will. The long bristles are used to assist in swimming. In some of these worms both the bristles and the podal spines are present, in others one or the other set of organs may be absent.

The spines, which with but few exceptions are present, are each gracefully curved like a long italic *S*, their shape resembling the line which artists have called the line of beauty. The free end, or the one projecting into the water, is forked in a way already described (Fig. 141). The body or shaft of the spine has, at some point of its length, a globular enlargement or a shoulder, below which the spine is often much narrowed. These organs are used by being protruded and forced against the surface over which the worm is travelling. They are arranged in a row on each side of the ventral surface, each row being composed of many clusters, and each cluster of from two to ten podal spines. The worm can protrude several clusters at once, or two on the opposite sides of the same segment or body-ring, but it seems unable to extend them irregularly.

The bristles are very flexible, and are arranged in two rows on the sides, near the upper surface, one row on each side. They are usually much longer than the width of the body, and may be so arranged that there are several or only one on each lateral margin of the segment. They are sometimes accompanied by a straight spine much shorter than the bristle, and pro-

jecting beside it. The free ends of these rudimentary
spines are occasionally finely forked. The bristles are
absent in some genera.

The worms are usually visible to the naked eye as
very fine whitish or yellowish threads, sometimes an
inch or more in length when extended. They are found
abundantly among aquatic plants, and in the mud of
shallow ponds. When allowed to remain in the collecting - bottle they will often make their way to the
lighted side, where some will form sheaths or protective cases from various floating fragments or particles.

The mouth may be close to the front end, or some
distance back, since in a few worms the front border is
extended as a long, flexible snout. The posterior border
is rounded in many forms, while in others it is expanded
into a broad, funnel-like region, with several finger-like
prominences surrounding it. In such worms these
parts are ciliated on the inside, the currents thus produced being supposed to bring at least a portion of the
oxygen needed for respiration. The alimentary canal
extends through the centre of the entire body, and is
usually crowded with the brownish remains of undi-
gested food. The whole cavity of the body outside of
the alimentary canal is filled with a colorless fluid visible only by means of the movements of the corpuscles
seen floating to and fro as the worm moves under the
cover-glass.

The beginner must not mistake this fluid for the
blood, which in many of the bristle-bearing forms is red

and contained in two distinct vessels, one extending lengthwise above, the other below the intestine. These vessels unite at both ends of the body, so as to form a long, closed tube, with branches springing from the front part, or from the upper or dorsal tube as it passes through each segment, where they then appear as pulsating loops. Usually the blood is impelled by the irregular pulsations of the dorsal vessel, a wave-like contraction passing along and driving the blood before it. In two genera (*Túbifex* and *Ocneródrilus*) there are little pulsating hearts attached to the dorsal vessel in the neighborhood of the frontal border.

Reproduction is by eggs or by transverse fission, the latter being most frequently observed.

Most of these worms live upon animal food, seeming to prefer Rhizopods and Rotifers to almost anything else ; a few are vegetarians.

Key to Genera of Microscopic, chiefly Fresh-water, Worms (Oligochæta).

1. Body with both bristles and podal spines (*a*).
2. Body with podal spines only (*b*).
3. Body with bristles only (*f*).
   a. Anterior extremity with a flexible, finger-like prolongation. *Pristina*, 1.
   a. Anterior extremity without a finger-like prolongation (*d*).
   b. Podal spines forked ; worms aquatic (*c*).
   b. Podal spines not forked ; worms aquatic (*g*).
    9*

b. Podal spines not forked; worms living beneath decaying bark of dead trees. *Enchytræus,* 2.

c. Podal spines, six to ten in each cluster, the clusters in two rows. *Chœtogáster,* 3.

c. Podal spines, two only in each cluster, the clusters in four rows. *Lumbrículus,* 4.

d. Posterior extremity without finger - like appendages (*e*).

d. Posterior extremity widened, ciliated, with several retractile finger-like appendages. *Déro,* 5.

d. Posterior extremity ciliated, with two long, non-retractile, finger-like appendages. *Aulóphorus,* 6.

e. Bristles and podal spines in separate rows (*h*).

e. Bristles and podal spines alternate in the same row. *Stréphuris,* 7.

f. Body variegated with brick red spots; blood colorless. *Æolosóma,* 8.

g. Podal spines in clusters of four each; body the color of raw meat. *Ocneródrilus,* 9.

g. Podal spines in clusters of two each. *Lumbrículus,* 4.

h. Worm with two small anterior pulsating hearts; blood bright red. *Túbifex,* 10.

h. Worm without a distinct heart; dorsal vessel pulsating; blood red. *Náis,* 11.

1. PRISTÍNA (Figs. 134, 135, 136).

Body nearly cylindrical, transparent, often very long, and showing that it is preparing to divide across the middle to form two worms. In these cases the proboscis of the new worm becomes conspicuous at the centre of the long body. The mouth is near the base of the snout-like prolongation, and this narrow extension of the upper lip varies much in length in the various species. The one represented in Fig. 134 belongs to a common form in the writer's locality, and is unusually long.

Fig. 134.— Snout of a Pristina.

The posterior extremity is commonly nearly as shown in Fig. 135, and surrounded by many short stiff hairs, it being the tail-end of the *Pristina* whose proboscis is shown in Fig. 134. Occasionally this part has three long trailing appendages, as in Fig. 136. The blood is usually red.

Fig. 135.—Posterior extremity of a Pristina.

The bristles are long and fine, and are often accompanied by one or more short, nearly straight, rudimentary spines. The podal spines are in two rows on the ventral surface, each cluster frequently containing as many as eight. The posterior part of the intestine is often ciliated.

Fig. 136.—Posterior extremity of a Pristina.

The worms are found among aquatic plants, seeming especially fond of Sphagnum and Lemna as a home.

2. ENCHYTRÆUS.

The body is white or yellowish white, thread-like and from about one-half to nearly one inch long. The worms are found under damp logs or beneath decaying bark, often in considerable numbers. The podal spines

are usually short, nearly straight, and not forked. The
blood is pale or colorless. There are two species, which
are not very difficult to distinguish from each other.

In one (*Enchytræus vermiculáris*) the body is yellowish white, and varies in length from five-twelfths to
eight-twelfths of an inch. The podal spines are in
clusters of from three to five spines each. This species
is usually found under damp and decaying logs, and is
less common than the next.

In the second (*E. sociális*) the body is opalescent white
and translucent, varying from five-twelfths to ten-
twelfths of an inch in length. The podal spines are
from five to seven in each cluster, the anterior fascicles
generally containing seven, the posterior five. The
mouth is triangular. This species is most frequently
found in more or less social groups beneath the moist
bark of old stumps or the decaying parts of trees, and
usually near the ground.

3. Chætogáster.

Body transparent, often showing evidences of transverse fission. The podal spines are in two rows, the
clusters containing four to eight spines each, being
usually most numerous towards the posterior extremity.
The mouth is large, oblique, and surrounded by many
very short stiff hairs. It is often used, when the worm
is on the slide, as a sucker, clinging to the glass and
drawing the body towards it. The intestine, in the
species common in the writer's vicinity, is much and

irregularly constricted, a feature which gives it the appearance of a series of various-sized pouches. The blood is very pale or colorless. The blood-vessels are distinct as narrow, pulsating, longitudinal tubes. *Chœtogáster* is one of the most interesting forms on account of its perfect transparency and the absence of bristles, which allows an uninterrupted view of the whole surface, as well as of the internal organs.

4. LUMBRÍCULUS.

The body is translucent, but often brightly colored at the sides or in the middle parts. The blood is bright red, and the dorsal vessel gives off several short, lateral, pulsating branches in each segment of the body. These short branches frequently approach the surface, and give it a mottled appearance, the spots fading and reappearing as the branches contract and expand. There are four rows of podal spines, with but two in each cluster. They are curved, forked at the end,* and have an enlargement or shoulder near their centre. At a short distance from the attached end of each pair there is often to be found another pair, which are small and may be overlooked on the front of the body when the worm is not dividing transversely. When it is undergoing transverse fission the posterior part may be so well supplied with these small secondary podal spines that their number and arrangement may confuse the

---

* Since this was written a species has been observed with undivided podal spines. It has been included in the Key.

beginner, the rows then appearing to be eight, with
two spines in each cluster, or four rows with clusters
of four spines each.   However, if the beginner will
examine the front half of the dividing worm, and be
guided by the podal spines there, he will have little
trouble in recognizing *Lumbrículus.*

5. Déro (Fig. 137).

The posterior extremity is broad and funnel-like, its
upper plane often being oblique.   It is finely ciliated,
as are the finger-like projections and the internal surface of the posterior part of the intestine, which is
connected with it and forms a portion of it.   The cilia
produce currents over these parts which are supposed
to absorb the oxygen for purposes of respiration.   The

Fig. 137.—Poste-
rior extremity
of a Déro.

finger-like processes vary in number from
two to eight.   They can be elongated or
drawn back into the funnel, which can also
be retracted and almost closed.   When extended they may be much longer than the
funnel-like termination of the body, or
they may not reach to its margins.   The blood is
red.   The podal spines vary from three to five in each
cluster.

These worms are often found on the sides of the collecting-bottle after it has been standing for some time.
They usually bury themselves in the mud, with the posterior part of the body and the expanded funnel-like region protruded from small mud-chimneys of their own

formation.   They may measure half an inch or more in length.

6. AULÓPHORUS (Fig. 138).

The posterior extremity is not wider than the width of the body, and the two finger-like appendages are straight or slightly curved.   They are blunt, and covered with short stiff hairs.   The worm usually builds a tubular sheath of various fragments and floating particles, in which it lives, but to the walls of which it is not adherent, as it frequently doubles on itself, glides through the tube, and thus reverses its position.   It moves by jerks, "alternately extending the fore part of the body and projecting the podal fascicles forward, and hooking into the surface on which it is creeping, and

Fig. 138.—Posterior extremity of an Auló-phorus.

then contracting the fore part of the body and dragging along the back part enclosed within the tube."* It often helps itself along by clinging to the slide by its protruded throat or pharynx.   The podal spines vary in number from five to nine in each cluster.   The fascicles of bristles are each accompanied by from one to three rudimentary spines, which are nearly straight, and end in a broadened, spade-like expansion.   The blood is red.

7. STRÉPHURIS (Fig. 139).

The podal spines and bristles are arranged alternately with each other, as in Fig. 139, and together form a single

---

* Dr. Leidy, in the *American Naturalist*, June, 1880.

row of clusters on each side of the lower surface of the body. The spines are slightly curved, long and forked, the bristles being three times their length. The mouth is triangular. The blood is bright red and the vessels large. The body is thread-like, transparent, and may be from one to two inches in length. The front end is whitish, the tail end yellowish. It lives in the mud beneath shallow water, and buries itself with about two-thirds of the tail end protruding and constantly vibrating. When disturbed it disappears into its burrow with astonishing rapidity. Dr. Leidy, who discovered this curious creature, says, "While walking in the outskirts of the city [Philadelphia] I noticed in a shallow ditch numerous reddish patches of from one to six inches square, which, supposing to be a species of alga, I stooped to procure some, when to my surprise I found them to consist of millions of the tails of *Stréphuris ágilis*, all in rapid movement. The least disturbance would cause a patch of six inches square so suddenly to disappear that it resembled the movement of a single body."

Fig. 139.—Podal spines and bristles of Stréphuris.

8. Æolosóma.

The bristles are of unequal length, and are arranged in clusters of four bristles each, the clusters forming a single row on both sides of the body. There are none in advance of the mouth, which is large U-shaped, the arms of the U pointing forward, the whole being surrounded with a thick border. The pharynx is broad and ciliated within. The body is colorless, the brick-red spots scattered over the internal surface giving the worm a beautiful appearance. *Æolosóma* is found in ditches among Algæ, on which it feeds. It is not very active in its movements. The blood is colorless.

Among the Sphagnum in the writer's locality there not uncommonly occurs a worm which I have ventured to identify as a member of this genus. It externally differs from the species referred to above in having fewer and larger red spots, which seem to be on the outer surface of the skin, where they are most abundantly collected near the two extremities, being fewest on the central region of the body. The bristles are so arranged that they appear to form two rows of clusters on each side, being separated into two groups in each cluster. The worm thus seems to have four rows instead of the two, as in the preceding species. Its movements are also much more active. It is also a vegetarian.

9. OCNERÓDRILUS.

This remarkable worm has thus far been found only in Fresno County, California, where it was obtained among fine Algæ growing to the sides of a submerged wooden box, and also occasionally in the mud, with a part of the tail end protruding and motionless. The body is rather less than an inch long, one-twelfth of an inch wide, and presenting the peculiar color mentioned in the Key. Its movements are very slow.

The podal spines are slightly curved, but not forked at the ends. They are arranged in clusters of four spines each, the clusters forming two rows, one on each side of the body.

The œsophagus is long and remarkably muscular. It is surrounded and somewhat obscured by a pair of large glands, and has near its posterior extremity two large appendages similar in structure to the œsophagus itself. The blood is yellowish-red. The dorsal vessel, at some distance behind the front end of the body, divides into three branches, which pass forward, and near the anterior border unite by means of a net-work of fine vessels. The worm has four hearts, two on each side of the dorsal vessel, one pair being near the eighth, and one pair near the ninth cluster of podal spines. The dorsal vessel divides in front of the first pair of hearts. The ventral blood-vessel is forked, but with only two branches.

10. TÚBIFEX.

A common and, in some places, a very abundant little worm, measuring from one-half to one and one-half inches in length. The body is thread-like in its narrowness, and is transparent and colorless, although the bright crimson blood gives it a hue so vivid to the naked eye that, where the worms are numerous, it often seems to tinge the mud in which they live. They are seldom found free-swimming, but live a comparatively sedentary life, with about one-half of the body buried in their burrow, the remaining parts protruding into the water, and constantly waving to and fro beyond the edge of the little tubular chimneys which they erect. These little towers are often conspicuous objects on the surface of the mud in shallow still water, the worms instantly disappearing into them at the slightest disturbance. Among certain French and German writers on the subject, there is a difference of opinion as to which end of the worm is buried and which end protrudes into the water. As the protruding parts are continually moving, and as the worms also dart into the mud with such astonishing swiftness, to decide the matter is rather difficult.

The bristles are comparatively short, and appear to be arranged in a single row on each side of the body, whereas there is really an additional row of podal spines on both sides of the worm. These podal spines are entirely retractile, and are therefore often overlooked unless specially searched for. Even then it will perhaps be necessary to compress the worm rather forcibly between the slide and the cover-glass before they will become conspicuous. They are forked, and but slightly curved.

With very high magnifying power (about eight hundred diameters) some of the bristles present a curious aspect. The free extremity is widened and forked, the two prongs of the fork being apparently connected by a thin membrane which is longitudinally striated. Sometimes this membrane splits into fine hairs. These widened bristles are most common on the young worms.

The bright red blood is contained in two principal tubular vessels, one above, the other below the tortuous intestine. The upper, or dorsal one, has connected with it near the anterior end of the body two little contractile hearts, one on each side, which can be seen through the hyaline animal throwing out the blood with considerable force. The two vessels are connected with each other by smaller ones, a pair in each segment or body-ring, one being on each side. There is also on each side of the body—two in each segment—a narrow colorless tube, ciliated within, and resembling those found in Nais and other aquatic worms. They are most conspicuous in the posterior rings, and are supposed to represent kidneys in function.

Tubifex is reproduced by eggs, which probably make their escape after the parents' death, and after the body has fallen to pieces, as the living creature has no passage for their exit. Huxley, however, says that they pass out through the segmental organs—the ciliated tubes just referred to.

11. NÁIS (Figs. 140, 141).

Body whitish or yellowish, usually very active. The

Fig. 140.—Náis.

podal spines and bristles are each arranged in a row on both sides of the worm, the bristles near the front end usually being longest. Each cluster of podal spines contains four or more. The mouth is round, the front border of the

body bearing numerous fine, short hairs. Two dark or
black eye-spots are generally present, one on each side
in advance of the mouth. The blood is red. Many
narrow, colorless tubes, with a ciliated lining,
are to be noticed on both sides of the intestine. They are much looped and twisted, and
are supposed to play some part in respiration,
or to represent the kidneys of animals higher
in the scale. They are bathed in the colorless fluid filling the cavity of the body, and
change their position rapidly as this fluid
flows to and fro, following the movements of the worm.

Fig. 141.
Podal Spine
of Náis.

Náis is the commonest of the aquatic worms, being
very frequently found among Algæ in shallow water, or
on the leaflets of various plants, especially, according to
the writer's experience, in Sphagnum, in company with
Pristina and Chætogaster.

Dr. Joseph Leidy's papers, published in the *Journal
of the Academy of Natural Sciences*, Philadelphia, and
elsewhere, are the only ones to which the student can
be referred for further information in connection with
the aquatic bristle-bearing worms, as Dr. Leidy is the
only naturalist who has seriously studied the American
forms and published the results of his work.

CHAPTER VIII.

RÓTIFERS.

WHEN these transparent microscopic animals are swimming or taking food, there is usually an appearance of two, small, rapidly rotating wheels on the front border of the body, an appearance that suggested the name of Rotífera, or Wheel-bearers, for the group. The two organs certainly do seem like rapidly revolving wheels when viewed under a low power, but they are in reality two disks or lobes surrounded by wreaths of fine cilia, which vibrate so quickly that the eye can perceive the effect only. It is by the action of these cilia that the Rotifer swims and captures food, the currents produced by them when the animal is at rest setting in towards the mouth, usually situated between the ciliary (or cilia-bearing) disks, and carrying particles of food which the Rotifer accepts or rejects. As a rule the ciliary disks are two separate organs, but they may be united into one, or the Rotifer may have the front margin of the body bordered by a single line of cilia, or the disks may be entirely absent, and replaced by long arms, as in *Stephanóceros*, or by clusters of long, fine hairs, as in *Floscularia*, both of which are Rotifers.

Most of these animals have eyes at some period of their life, or little red specks supposed to be imperfect

eyes. They are often to be noticed near the front of
the body in young individuals, but in the old they are
as often absent. Their number and position are sometimes used as characters by which the genera and species are classified, but, since they disappear with age,
they cannot be of much value for this purpose, certainly
not to the beginner.

The body is inclined to be cylindrical, yet there are
some resembling flat disks and oblong figures. Neither
are they all free-swimming. Some are permanently adherent to the leaflets of aquatic plants or other submerged objects, but these generally form a protective
sheath about themselves, into which they retire when
frightened or disturbed, in a manner similar to that of
some Infusoria; and, as in the Infusorial loricæ, the
sheaths may be formed of a stiff membrane, or of the
softest and most gelatinous material, or they may be
built of particles of dirt or rejected food fragments. In
all instances the sheaths are the work of the Rotifers
inhabiting them, and none of the sheath-building Rotifers are free-swimmers.* Most of the free swimmers,
however, may become temporarily adherent by means of
their foot and toes. The body of these free-swimming
forms may be soft and flexible, and without any greater
protection than is afforded by the skin, or it may be enclosed within a hard, shell-like coating called a carapace.
The bodies of all the sheath-building Rotifers are with-

* Since the above was written three have been discovered in England. But this need not trouble the beginner.

out a carapace, the lorica being a sufficient protection. In the other kinds the carapace is colorless and transparent as a glass box, all the creature's organs being plainly visible through its walls. The front part of the body, which bears the cilia or the ciliary disks, and often the long tail-like prolongation of the posterior part, can be drawn within the carapace, and the Rotifer thus shut in and protected from harm. The soft-bodied forms have a similar habit of drawing in the two ends, taking advantage of the hardened skin. This is one of the Rotifer's characteristics.

The long tail-like part at the posterior end of the body is called the foot, and the two or more short divisions at the free end of the foot are, of course, the toes. The true tail of the Rotifer is usually a small affair, which the beginner must not mistake for the more important foot, although it is placed on the foot, sometimes quite near the body. It may be represented by a single short point, it may be in two parts and more conspicuous, or it may be entirely absent. The uses of the foot seem to be to act as a rudder to guide the Rotifers when swimming, as they do in a hurried, headlong way, and also to anchor them when they desire to fish for food. The toes then adhere to the surface of the slide or of any other object, anchoring and holding the animal against the propelling power of the ciliary disks. In some of the group, especially in the commonest of all—*Rotifer vulgáris*—the whole foot is arranged with joints that slide on each other like the joints of a spyglass. In this and similar forms the Rotifer can not only swim, but it can crawl by fixing the front of the body against the slide, drawing in the telescopic joints of the foot and clinging with the toes; the front is then loosened, the foot extends and carries the whole body forward for a short distance, when the action is repeated. A Rotifer can do this with surprising rapidity, and so travel over considerable distances in a short time.

The mouth is usually placed between the two ciliary disks, when they are present, near the centre of the frontal portion of the body, or, in some forms, it is placed near the front, but on the lower surface of the animal. Those with the mouth in the last-mentioned position usually feed by gliding along with the front of the body in contact with the plant, tearing and biting off small particles as they go. These may be called the nibbling Rotifers. Following the mouth there is often a tubular passage leading to a pair of wonderful jaws inside of the body, which, with a low-power objective, can be seen in action through the transparent tissues of the Rotifer. These jaws are always present in these creatures, and are a great help to the beginner, for as soon as he observes them pounding and crunching away inside of a transparent, legless, microscopic animal, he may be sure that his specimen is a female Rotifer. The ciliary disk may be absent, or replaced by arms, hairs, or some other substitute, but if these internal jaws are present the specimen is a Rotifer, and can be nothing else. By some observers these curious organs are called

10

the gizzard, which they are not. The best word to apply to them is *mastax*.

The mastax is the most hard-working part of the creature's anatomy, except, perhaps, the cilia. When the currents produced by the latter bring an acceptable morsel of food to the mouth, it is passed down to the mastax, where it is crushed and allowed to go on to the stomach. In some Rotifers this part is very complicated. In the simpler forms it consists of two apparently semicircular plates surrounded by a thick envelope of powerful muscle, the flattened sides acting against each other and crushing the food between them. The surface of each plate very often bears several transverse parallel ridges, to be seen with a high power, each ridge projecting a short distance beyond the straight internal edge, to form short teeth. These ridges, when the mastax is closed, are received in the depressions between those on the opposite plate, thus making an effectual crushing instrument. In other forms the mastax consists of three parts, one being immovable, and used as an anvil on which the other two pound the food as it passes by. In the nibbling Rotifers the entire mastax is protruded through the mouth, and bites, tears, and nibbles at acceptable food masses.

If the beginner finds it difficult to make out the form and structure of the mastax, as he probably will when it is examined in action within the body, he may succeed by killing the Rotifer with a strong solution of caustic potassa allowed to run under the cover-glass — a small drop at a time. This will dissolve the soft parts, and permit the hard, insoluble mastax to float out, when it can be examined with a high-power objective.

The Rotifers are reproduced by eggs, which are sometimes hatched within the parent's body, when they are said to be ovo-viviparous. This, however, is not common. The eggs are usually semitransparent, ovoid bodies, very often to be seen on the slide among other matters, with the Rotifer partially developed, and the mastax grinding away inside of the unhatched body where it cannot possibly have anything to crush. The only parallel to this of which I know is Professor Agassiz's statement that the jaws of the young snapping-turtle snap while the creature is still in the egg. The Rotifers may drop their eggs anywhere and leave them to the care of Nature, or they may prudently attach them to a leaf or some other aquatic object. Very often they are adherent to the posterior part of the parent, and are carried about until the young are hatched. In those permanently attached Rotifers that form a soft sheath this is a common occurrence, and several eggs may at almost any time be seen in the lower part of the lorica, or fastened to the animal's foot. In such instances, when the young are hatched they creep up between the parent's body and the side of the sheath and escape at the front. They swim about for a short time, and then secrete or build a sheath of their own, which they never voluntarily leave. The eggs are usually smooth ; some-

times, however, they are covered with short spines or
hairs.

It is a curious fact that although there are male and
female Rotifers, the males are seldom seen. In some
species they have never been found, and are therefore
entirely unknown. Those that have been discovered
are much smaller than the females of the same species.
They are always free-swimming, and are without a mastax and alimentary canal, or with the latter so imperfect
that it is useless. Male Rotifers, therefore, never take
food. It is not probable that the beginner will meet
with them, or at least will recognize them as the
males.

This group of animals is almost as common and abundant as the Infusoria, and they are found in similar
places. They are specially fond of hiding in masses of
Ceratophyllum. Indeed, almost any pond or shallow
body of still water may be examined with a certainty
of finding them. They have even been sparingly obtained from the moss that grows between the bricks in
damp pavements. Some species develop in vegetable
infusions, but as a rule they prefer fresh water. The
beginner will, of course, not expect that all the genera
and species will be included in this little book. He will
obtain very many whose names he cannot hope to learn.
He can, however, know them to be Rotifers by the presence of the mastax, which makes them one of the most
easily recognizable groups of microscopic animals. They
form an interesting class of creatures for microscopic

study. Very few of our American forms have been
investigated, and there is no one book in the English,
nor, so far as I know, in any other language, to which
the beginner can be directed for help. The American
" Wheel-bearers " form a large field which ought to be
cultivated. There is room for many discoveries, and
an opportunity to greatly add to the world's store of
scientific information.

In using the following Key, the beginner must remember that the sheath of some of the Rotifers is very
often colorless and rather difficult to see clearly, unless it has particles of dirt or other matters adherent
to it. At other times it may be conspicuous. The
Key refers to only those forms included in this book.

Key to some Genera of Rotifers.

1. In a gelatinous or other kind of sheath (*a*).
2. Not in a sheath, but growing in attached clusters (*d*).
2. Free-swimming (*e*).

   a. Clustered; sheath soft, gelatinous, colorless. *La-cinulária*, 1.

   a. Not clustered; sheath gelatinous (*b*).

   a. Not clustered; sheath not gelatinous (*c*).

   b. With five long, erect, ciliated arms on the front
border. *Stephanóceros*, 2.

   b. With five clusters of many long, fine, radiating
hairs on the front border. *Flosculária*, 3.

   b. With two ciliary disks; sheath tubular, often
branched. *Actinúrus*, 4.

   c. Sheath formed of rounded brownish pellets. *Melicérta*, 5.

   c. Sheath membranous, brownish or colorless. *Límnias*, 6.

   d. Ciliary disk horse-shoe shaped. *Megalótrocha*, 7.

   e. With carapace (*g*).

   e. Without carapace; ciliary disks two (*f*).

   f. With ten or twelve short, scattered, recurved
spines; toes three. *Philodína*, 14.

   f. Without spines; foot with telescopic joints, toes
two. *Rótifer*, 8.

   f. Without spines; foot with telescopic joints, toes
three, the middle one longest. *Actinúrus*, 4.

   g. Carapace with a vizor-like projection in front.
Stéphanops, 9.

   g. Carapace circular; foot long, cylindrical, retractile.
Pterodína, 10.

   g. Carapace vase-shaped; foot long, with two very
long toes. *Dinócharis*, 11.

   g. Carapace with six long, narrow, movable fins on
each side. *Polyárthra*, 12.

   g. Carapace with several tooth-like spines on the front
border. *Brachiónus*, 13.

1. LACINULÁRIA.

The clusters contain numerous individuals secreting a common, soft, colorless, or pale yellowish short sheath without any special shape; it surrounds only the posterior part of the colony and can serve as a very slight protection, if any. The Rotifers are somewhat trumpet-shaped when extended, and to a certain extent resemble Megalótrocha (Fig. 147). The ciliary disk is single and horseshoe shaped. It is closed and drawn partly into the body when the Rotifers retire into their apology for a sheath, as they often do, the whole colony continually waving and bobbing and bowing as the members retire, or ascend and expand themselves. The sheaths usually form a little mass of jelly-like substance, from all parts of which the Rotifers project. The colonies are commonly adherent to Ceratophyllum or Myriophyllum.

2. STEPHANÓCEROS (Fig. 142).

The body of this the most beautiful of all the Rotifers is somewhat spindle-shaped. It ends in a long, flexible, tail-like foot which is attached to some submerged object, and has five long, slightly curved arms arranged in a row about the edge of the front border. These arms are held aloft and form a most effectual trap for wandering Infusoria, which are attracted or drawn into it by some means not easy to make out. The front of the body is like a deep open funnel leading down to the mouth, mastax, and stomach. The ordinary ciliary disk is absent, being re-

Fig. 142.
Stephanó-
ceros.

placed by the arms, but around the inside border of the funnel-like front there seem to be many fine cilia which may produce the currents in the water.

They are very difficult to see even with a high - power objective.

The sheath is usually colorless and transparent, with considerable firmness. It often surrounds the body up to the origin of the arms.

When a small animal once enters the cage formed by the arms it seldom escapes, but is gradually driven down into the funnel, when the Rotifer partially closes the front opening, and with a very perceptible gulp swallows and passes it on to the mastax.

Stephanóceros does not seem to be very common. The writer has found it sparingly on Myriophyllum as late as the middle of November.

3. FLOSCULÁRIA (Fig. 143).

The front of the body is here also like an open funnel, the narrow part leading to the mastax. The ciliary disk is replaced by five little rounded elevations on the front margin, each bearing a thick cluster of long, fine, radiating hairs, which are flexible, and movable at the animal's will, but which never vibrate like cilia. The long foot is attached to a submerged object, and is surrounded by a soft and transparent sheath. When the Rotifer retires into this protective covering, it folds the wide front part of the body together, the clusters of long hairs seem to become much tangled into a single bunch, and

Fig. 143.
Flosculária ornáta.

the creature slips back into the sheath. When she comes out, the bunch of hairs tremble in a very pretty way, reminding the observer of the quivering appearance of hot air often seen on a summer day. The front border opens, the clusters of hairs are spread apart, and the Rotifer is ready for something to eat. Any little animal slipping in between the hairs seldom comes out again. The Floscularia gently contracts the frontal opening and directs the victim towards the mouth, where it is gulped down as in Stephanoceros, and the mastax finishes it. Several eggs are often seen attached to the foot. This splendid Rotifer is common, and where one is found others will usually be near by.

4. ACTINÚRUS (Fig. 144).

When a bottle of pond water and various plants is allowed to stand for a while undisturbed, there will often form on the sides very delicate, thread-like objects, frequently branching and otherwise resembling brownish Algæ, waving and trembling as the bottle is stirred. They are so soft that they can hardly be removed with the dipping-tube without breaking them. They are the sheaths of a Rotifer. She makes them from a sticky secretion exuded by her body, and small particles of any matters that may be floating in the vicinity. The inside

Fig. 144.
Actinúrus.

seems to be smooth, but the outside is rough and irregular. The Rotifer projects from the open end, clinging

10*

to one inside wall by the tips of her toes, and as the
tube lengthens by the deposit of new material at the
top, she takes a step forward so as to keep her expanded
ciliary disks in the open water.   If the student will allow a mixture of fine indigo and water to run under the
cover-glass, he will see the formation of the sheath.   A
blue ring of indigo will very soon appear at the top of
the soft tube.

In appearance the Rotifer resembles *Rótifer vulgáris*
(Fig. 148), and when out of the tube, which she can
leave at will, has similar actions.   There are two ciliary
disks, and usually two eye-spots.   The foot is long, and
can be drawn into the body by telescopic joints.   It has
three toes, the middle one being the longest.   The eggs
are hatched within the parent's body.   The Rotifers occupying the branches of the sheath are probably all
members of the same family — mother, children, and
grandchildren—the young forming the branches.

There is a species of this genus which does not form
a sheath.   It may be known by its resemblance to Rotifer vulgaris, and by the three toes, the middle one
being the longest.   .

5. Melicérta (Fig. 145).

The sheath of *Melicérta* resembles that of no other
common Rotifer.   It is built of pellets, which she makes
and places in rows around her body, thus erecting a reddish or yellowish-brown lorica that cannot be mistaken.
The body itself is colorless, and is always attached by

the tip of the long foot to an aquatic object. The ciliary disk consists of four parts or lobes of different
shapes and sizes, and the little creature also has a very peculiar and rather complicated organ for making
the pellets. The whole front part of the body
can be folded together into a rounded mass
when Melicerta is frightened and retires to
her sheath. When her fright is over, she
slowly protrudes this rounded mass from the
aperture, gradually spreads it open, sets the
cilia at work, and proceeds to eat and build.
The last she seems to do almost continuously.

Fig. 145.
Melicérta
riugens.

As her body grows, her house must be enlarged to receive it.

The ciliary disk of Melicerta will repay the most careful study. And careful observation will be needed to
learn just how the three distinct currents that she makes
in the water are produced. One current brings food
particles to the mouth, where she selects the acceptable
morsels and passes them on to the mastax; a second
current carries away the fragments for which she has no
use; and the third sets in towards the little organ that
makes the pellets. This is a small cavity into which
the building material is poured, and where it is turned
about rapidly by the fine cilia which line it. A sticky
secretion is exuded that causes the particles to adhere to
each other, and the revolving motion gives the pellet
the shape of a Minié-bullet. When the latter is formed
to the Rotifer's liking, and all is ready for the final act,

Melicerta turns herself in the tube, bends her body, and deposits the pellet on the top row, where she cements it in place with an invisible insoluble cement. The whole is done so quickly that the first time the observer sees it he is so surprised that he sees nothing. It is remarkable that, as a rule, she forms the pellet while standing on the side of the sheath opposite to the point where she means to place it. The pellets have often been described as round balls, but the student will see that they are conical, and that the pointed ends are on the outside. Melicerta is quite common on Ceratophyllum.

6. Límnias (Fig. 146).

The sheath that this Rotifer forms is a rather stiff, membranous, nearly cylindrical tube, somewhat widest at the upper part. When young it is usually colorless and smooth, but it changes with age, becoming brown or blackish, and floating particles roughen it by adhering to the outside. The animal living within it is colorless, and has the ciliary disk divided into two lobes, which she folds together when frightened and forced to retire to the back part of the sheath for protection. The sheath is secreted from the body of the Rotifer; it is not built of particles picked out of the currents from the ciliary disks. It is common on the leaflets of Ceratophyllum, and is probably named *Límnias ceratophýlli* for that reason, although it is found almost as often on Myriophyllum.

Fig. 146. Limnias ceratophylli.

There is another species of Limnias also quite common in the writer's locality, which differs from Limnias ceratophylli in having the sheath apparently formed of narrow rings, so that the edges, as seen under the microscope, seem finely waved or scalloped. By this it can be easily distinguished from the above. It is named *Límnias annuláta.*

7. MEGALÓTROCHA (Fig. 147).

The clusters formed by *Megalótrocha* are sometimes so large that they are visible to the naked eye as whitish bodies clinging to Myriophyllum, which it seems to prefer. With a pocket-lens the individual Rotifers may be seen rising and bobbing as they expand or contract, but a low power of the compound microscope is needed to appreciate their beauty. The expanded body is rather trumpet-shaped, very soft and flexible, and when young is colorless. As it grows old it becomes slightly yellowish. The eggs are often to be noticed adhering to the lower

Fig. 147.
Megalótrocha.

part of the parent. When the young one is hatched it either remains in the old colony or it leaves and founds a new cluster, so that in favorite localities colonies of almost any number of members may be obtained. The Rotifers of old colonies are often infested by an Infusorial parasite, which runs over the surface and apparently feeds on the mucous matters secreted by the Rotifers' skin. It is called *Chilodon megalo-*

trochæ, and somewhat resembles the Chilodon shown in Fig. 128.

8. Rótifer vulgáris (Fig. 148).

This is the commonest of all the Rotifers. The body is spindle-shaped, tapering to both ends when the two ciliary disks are unfolded. The foot has two short toes, and can be drawn into the body by its telescopic joints. Between the two ciliary lobes is a cylindrical projection ciliated on the tip, and nearly always bearing two little red eye-spots close together. It is called the proboscis. When hungry the Rotifer clings to the slide by her two toes, expands the ciliary disks, and sends a food-bearing current through the mouth to the mastax. When desirous of changing her place, she may either loosen her hold with the toes and be carried through the water by the action of the cilia, or she may fold the ciliary lobes together and go looping about by clinging with the tip of the proboscis while she draws up the foot, when, fastening it to a new point, she lets go with her proboscis, extends the body, takes a new hold with the foot, and thus moves about quite rapidly, somewhat after the manner of the "measuring-worms."

Fig. 148. Rótifer vulgária.

9. Stéphanops (Fig. 149).

There are several species of these pretty little Rotifers, all of which may be known as members of this genus by the extension of the front of the carapace over the ciliary disk, like the visor of a boy's cap. A not uncommon species is shown by Fig. 149, in side view, so as to exhibit the long, movable bristle springing from the back, and the curved visor which, in the figure, looks like a line above the frontal cilia. The Rotifer

Fig. 149.—Stéphanops.

is one of the nibblers. The mastax is protruded from the mouth, which is near the front of the lower flattened surface, and bites and tears the food it meets with. It is often to be seen gliding over aquatic plants, nibbling as it goes. The carapace is thin and quite flexible. It extends over the sides of the body, so as to give the Rotifer an ovate outline when seen from above or below. The bristle on the back is very movable and flexible.

In one of the species the carapace is prolonged at the posterior border into two lateral teeth. In another this part is without teeth, and the dorsal bristle is also absent, as it is in all the known American species, except the one shown in Fig. 149.

10. PTERODÍNA (Fig. 150).

The carapace is almost circular, much flattened, and perfectly transparent. The anterior border has a broad notch with rounded margins, over which extends a lip with a central rounded projection. The ciliary disks are two, and rather widely separated. In the figure they are shown retracted into the body. There are usually

two eye-spots. The foot is long, tail-like, very flexible,
and apparently formed of narrow rings. It
can be withdrawn entirely into the carapace,
and the Rotifer seems to take pleasure in
doing so. It has no tail, but is terminated
by a small sucker bordered by a circle of
fine cilia. The Rotifer is often seen among
Ceratophyllum leaflets.

Fig. 150.
Pterodina.

11. DINÓCHARIS (Fig. 151).

The transparent, glassy carapace is rather squarely
vase-shaped and somewhat flattened. It generally has a tooth-like projection on each side
of the posterior border. The Rotifer can always be recognized by its two very long toes,
in one common species one toe being very
much longer than the other. The foot is
formed of two joints slightly enlarged at the
ends.

Fig. 151.
Dinócharis.

12. POLYÁRTHRA (Fig. 152).

The form of the carapace is somewhat like an egg,
with both ends cut squarely off. The character, however, by which the Rotifer may always be
known is the presence of the twelve long,
serrated fins which project backward from
the front part of the upper and lower surfaces. They are arranged in clusters of
three fins each, one cluster being on each
side below, and one on each side above. By them the

Fig. 152.
Polyárthra.

Rotifer makes long, quick, very sudden leaps, often jumping so rapidly that it can hardly be seen; it appears to spread the fins and disappear. Occasionally it turns a complete somersault. The cilia are arranged in a row along the front border. There is no foot. The mastax is pear-shaped and large, but its structure is difficult to make out. The Rotifer has only one eye, which is near the centre of the upper front surface. The little creature has been called by some writers the "sword-bearer," and is said to be quite common in some localities, but I have never been fortunate enough to find it.

13. Brachiónus (Fig. 153).

There are several species of this genus, all of which may be known by the presence of a carapace with several tooth-like projections or spines at the front, and often also at the rear, by the two ciliary disks and the single eye-spot. The species whose empty carapace is shown in Fig. 153 is strictly American. It is very attractive in its glass-like transparency, active movements, and beautiful carapace. It was discovered by Mr. H. F. Atwood, of Rochester, and named *Brachiónus cónium* by him.* It is quite common, and may be easily recognized by the ten long teeth or spines on the front border—the central one on the upper side or back being largest, and bent at a right angle—and by

Fig. 153.
Brachiónus.

* *American Monthly Microscopical Journal*, June, 1881.

the four posterior ones. The foot is long and narrow, and has two toes. Eggs are occasionally to be noticed attached to the posterior part of the carapace.

14. PHILODÍNA (Fig. 154).

This is readily distinguished from the common Rotifer (Rotifer vulgaris) by the spines scattered over the back and sides of the hardened and minutely roughened body, by the three toes, and by the two eyes being some distance from the front border, while in Rotifer vulgaris they are close together on the proboscis. The species referred to is shown in the figure with the body partly contracted, the ciliary disks and foot entirely so.

Fig. 154.
Philodina.

The body is flexible, yet the skin is hardened and bears the conspicuous recurved spines, two of which are on the sides—one on each—and pointing forward. The tail is divided into two parts, which are shown in the figure, projecting beyond the body. The Rotifer is peculiar, and not uncommon. I have found it in summer, and have taken it from under the ice in February.

CHAPTER IX.

FRESH-WATER POLYZOA.

THE reader now approaches a group of microscopic animals whose beauty is so exquisite, so delicate, so refined in its comeliness and grace, that no description could be too extravagant, no rhetoric too fervid when applied to the charming little creatures. Yet most of this fairness seems wasted so far as human appreciation is concerned, for how few among the millions of human beings in all the land know, or care to know, what the *Polyzoa* are, or how they look, or where they live, or whether they live at all? Nature was never in better mood than when she began the development of the Polyzoa, so she fashioned them with care, and placed them most abundantly in all our slow streams and shallow ponds, where they live and die and melt away in the shade of the lily-leaves, where no human eye sees their loveliness until a wondering lover of Nature spies them and is happy.

The word Polyzoa is formed of two Greek words meaning "many animals," referring to their habit of living in colonies which sometimes reach an immense size. They are, with but one exception, always attached to some submerged object, except immediately after leaving the egg, when the young animal leads a short, free-swimming life. When once attached they are adherent till death. The animals themselves are small, but often apparent to a trained eye; they are always visible under a good pocket-lens. The colonies, however, of all the fresh-water forms need no magnifying; some of them are very conspicuous. These communities are formed of the protective coverings or sheaths secreted by the animals. Some take the form of very narrow, brownish tubes, adherent to the lower surface of floating chips, boards, waterlogged sticks, or even occasionally to lily-leaves or the submerged stems of grasses. The little tubes branch like miniature trees, and spread over the surface as if the delicate tree had been flattened down and pressed so hard that it could never again rise up; or they may be attached by the base only, the trunk and the branches then floating and waving in the water. The animals secreting these tubes live in them, projecting a part of the body beyond the orifice, and very quickly retreating when frightened. And they are usually very timid, retiring into the tubular home at a slight disturbance of the water, needing a long time in which to recover and again look out at the entrance and spread their beautiful tentacles.

In other forms the colony is surrounded by a thick, rather firm, jelly-like material, from which the animals protrude themselves, and into which they retreat. These jelly masses are usually colorless and semitransparent, or they may be tinged a pale red. They are to be found in the purest of still water, adherent to sticks, capping a

submerged stump with a cushion of living jelly, clinging like crystalline globules to any projecting rootlet
or water-soaked object beneath the surface, even to
smooth stones. In bulk they may be like a boy's marble, or a cart-wheel, with every intermediate size. They
vary so much that to find a good comparison is not
easy, and it is only right to say, lest some lover of these
lovely creatures should be envious, that a colony the
size of a cart-wheel has, in the writer's locality, been
found but once, the foundation of this remarkable
growth probably being the rim of an old wheel.

When the tubular or the jelly-like colonies are removed to the collecting-bottle, they appear lifeless and
unattractive. The jelly may excite wonder by its size,
or curiosity to know what it can be, yet otherwise it
will not be noticeable. But wait a while. Place the
bottle in the shade and wait a few minutes; then
with a pocket-lens look at the surface of the jelly or
the tips of the branching tubes. Treat them with
care; move them gently. The little creatures are
easily frightened, and like a flash leap back into their
protective case. Perhaps while you gaze at the reddish
jelly a pink little projection appears within the field of
your lens, and slowly lengthens and broadens, retreating
and reappearing it may be many times, but finally, after
much hesitation, it suddenly seems to burst into bloom.
A narrow body, so deeply red that it is often almost
crimson, lifts above the jelly a crescentic disk ornamented with two rows of long tentacles that seem as

fine as hairs, and they glisten and sparkle like lines of crystal as they wave and float and twist the delicate threads beneath your wondering gaze. Then, while you scarcely breathe, for fear the lovely vision will fade, another and another spreads its disk and waves its silvery tentacles, until the whole surface of that ugly jelly mass blooms like a garden in Paradise—blooms not with motionless perianths, but with living animals, the most exquisite that God has allowed to develop in our sweet waters. Perhaps you make an inarticulate cry to your companion, who is probably wondering why you are so still and what you are doing on the ground with the lens so close to the bottle, and as he too gets down and brings his lens to bear, maybe he jars the water, and the lovely Polyzoa flash their tentacles together and dart backward into the mass, leaving it as indescribably ugly as before. If he brings you to task, tell him to wait and look. And while he looks the little bodies again slip outward, the crescentic disks again spread wide open, the shining tentacles unfold and curl and lash the water until once more the ugly jelly mass becomes a thing of indescribable beauty. This is *Pectinatélla*, well named the magnificent.

The jelly is formed by the animals, and is in reality a collection of protective loricæ, the huge masses often found being the result of the increase in the numbers of the Polyzoa inhabiting them ; or, as must frequently occur where they are very abundant, of the union of many contiguous growing colonies. A single animal begins the colony; it becomes two by a process of budding, the bud finally becoming another Polyzoön, secreting more jelly, budding in its turn, so that the community may in the end contain numberless members, and the mass may measure several feet in diameter. The color of the animals is usually a pale red or flesh tint, deepening to crimson about the mouth, which is placed near the centre of the crescentic or horseshoe-shaped disk of tentacles. In the largest, and therefore the oldest, colonies the jelly may exhibit many scattered white spots composed of carbonate of lime.

There is another jelly-forming colony called *Cristatélla*, which the beginner may mistake for young Pectinatella. It is to be distinguished by the absence of those great masses which characterize Pectinatella, by the general appearance of the colony, and by its motion. A community of Cristatella is usually long and narrow, often measuring several inches in length. One species is about eight inches long, one-fourth of an inch wide, and one-eighth thick. Young colonies are, of course, smaller, and are rounded. It has the power which no other fresh-water Polyzoön possesses — to travel from place to place. It moves very slowly, a colony about an inch in length moving an inch in twenty-four hours.

All the fresh-water Polyzoa, of which there are several genera and species, have on the front part of the body a disk which bears the tentacles. It is named the lophophore, and is, in some forms, horseshoe-shaped, in others nearly circular. The tentacles are arranged on it as on a base, usually in a double row. The word is Greek, and means "wearing a crest."

In those Polyzoa which secrete hardened, tubular, tree-like sheaths on the surface of submerged objects, the lophophore is protruded from the orifice in the end of the branch much as in Pectinatella, and there is only one animal to each limb and hollow twig. The protrusion and expansion of the lophophore can be seen with a pocket-lens, as in Fig. 156 (from Professor Alpheus Hyatt's work on the Polyzoa), when it resembles in form that of Pectinatella. Those inhabiting the tubular sheaths seem much more timid than the gelatinous forms, retreating on slighter provocation, and remaining longer before they reappear and again spread the lophophore and tentacles. They are quite as graceful and attractive — perhaps they are more so, since they seem more delicate and less able to protect themselves.

The tentacles are finely ciliated, as the microscope will show. The currents produced by the active vibrations of the cilia on the sixty to eighty tentacles of Pectinatella, or the eighteen to twenty in other members of the group, are quite powerful, and setting in towards the centre of the lophophore, they sweep the entrapped food to the mouth. The body of the Polyzoön is a transparent, membranous sack, with the lophophore and the mouth at the free end, most of the rest being immersed in the jelly or concealed in the brown opaque sheath. The mouth has on one border a short, tongue-like organ, which can close the opening and prevent the escape of the food. Extending from the mouth to the stomach is the food passage or œsophagus, the stomach itself being a widened tube, usually conspicuous on account of its contents and the alternate narrow, reddish-brown and yellow bands traversing it lengthwise. It is suspended in the hollow body, and is bathed by a colorless fluid which fills the cavity and extends also into the hollow tentacles. The stomach is followed by the tubular intestine, which curves forward, and generally opens below and on the outside of the lophophore. The animals have no heart and no blood, unless the liquid in the space between the outer walls of the stomach and the walls of the body can be said to be blood.

When the animal is frightened, the sides of the lophophore close together, the tentacles collect themselves into a bundle, and the front of the Polyzoön is drawn back into the body, and a muscle around the border closes that opening. The jelly of Pectinatella and the hardened tubes of the other forms are, therefore, the protectors of the body, while the body receives and encloses the lophophore and tentacles, which are thus doubly protected. When the danger is past, the tips of the bundle of tentacles are very cautiously pushed out into the water, the lophophore follows, and if the creature's confidence is restored, the crowns are spread open in all their indescribable grace and beauty.

The favorite food consists of small Algæ and Infusoria, which the ciliary currents sweep towards the mouth, the tentacles forming a cage from which the little ani-

11

mals seldom escape unless the captor is willing. And
not only are the tentacles used to capture the food, but
" for a multitude of other offices. They are each capable of independent motion, and may be twisted or
turned in any direction; bending inward, they take up
and discard objectionable matter, or push down into the
stomach and clear the œsophagus of food too small to be
acted on by the parietal muscles."

To examine the Polyzoa under the microscope demands a very deep cell to hold a large quantity of water
and to prevent the cover-glass from pressing on the
bodies. It is often better to place the microscope in an
upright position and omit the thin cover. In this arrangement the water trembles easily, and not only inter-
feres with the distinctness of the image, but terrifies the
timid creatures on the slide. The observer must, therefore, be careful not to touch the table, and to make his
examination in a quiet room. They will ask a little attention and some gentle treatment, but what they will
show with the help of a one-inch objective will amply
repay the outlay of time and patience.

The following Key to the genera will help the student to name the forms he may find.

Key to Genera of Fresh-water Polyzoa.

1. In a jelly mass (*a*).
2. In adherent, branching cylindrical tubes (*b*).
3. In adherent, branching colonies formed of tubular,
   club-shaped cells (*c*).

4. In adherent, pendent stems formed of urn-shaped cells (*d*).

    a. Jelly mass rounded, adherent, often very large. *Pectinatélla*, 1.

    a. Jelly mass lóng, narrow, slowly travelling. *Cristatélla*, 2.

    b. Lophophore horse-shoe shaped. *Plumatélla*, 3.

    b. Lophophore circular. *Fredericélla*, 4.

    c. Lophophore circular, tentacles in a single row. *Paludicélla*, 5.

    d. Lophophore circular or oval, tentacles in a single row. *Urnatélla*, 6.

1. PECTINATÉLLA MAGNÍFICA (Figs. 155, 155*a*).

The appearance to the naked eye of the colorless jelly-like substance surrounding the bodies, and of the animals themselves, has already been referred to.

Pectinatélla is not sensitive to sound, but a jar or shock to the water sends the animals into their contracted state very suddenly. The colonies are numerous throughout the summer and until October. They are most frequently found in the shade, although they may live in the sun if below the water. Exposure to air and sunlight together is speedily

Fig. 155.—Pectinatélla magnifica.

fatal. Therefore transfer the jelly to the collecting-bottle as soon as possible, otherwise you will have, on

your return home, nothing but a softening, slimy mass
that will soon force you to throw it away. If suspended
in a large vessel of water kept very fresh by frequent
change, Pectinatella will live for some time in captivity.
In Fig. 155 (after Hyatt) is shown a small colony with
the lophophores and tentacles expanded and enlarged, as
they appear with a good pocket-lens. The absence of
color and motion, however, makes a great difference in
their beauty.

In old colonies, especially late in the season, there are
often to be seen very many small, rounded, brown
bodies, which, as the animals die, float to the surface of
the water. These are the winter eggs or statoblasts.
They are formed within the body, and escape only when
the Polyzoön dies and melts away, when they float out
and remain unchanged until the warmth of spring develops them. Under the microscope the stat-

Fig. 155a.—
Statoblast
of Pecti-
natélla.

oblasts of Pectinatella are seen to be encircled
by a row of double hooks, as shown in Fig.
155a. I have collected them late in the fall,
and, keeping them in a small aquarium in a
warm room, have had them hatch out in November. The young fastened themselves to the sides
of the glass bowl, where they appeared like delicate
grains of translucent pearl. There was no jelly at this
early stage, and each little Pectinatella stood alone, consequently all the internal organs were even more distinctly visible than usual through their hyaline bodies.
I hoped to see them develop into colonies, but the sur-

roundings were not quite favorable, perhaps the proper
food was not attainable, so they died.

2. CRISTATÉLLA.

The form and movements of *Cristatélla* have already
been referred to on page 225. The young colonies are
rounded, and are found in the same localities with Pectinatella. The statoblasts are circular and have two
rows of double hooks, one row around the border, the
other nearer the centre. In both, the hooks are not simple as in Pectinatella, but have several branches at the
top of the stem, and the tips are forked.

According to the writer's experience, Cristatella is not
common.

3. PLUMATÉLLA (Fig. 156).

The tubes containing the animals may be attached
only at the base, or the whole colony may be adherent
to the submerged surface on which it grows. It is to
be found in shallow water,
usually near the shore. To
see the lophophore and expanded tentacles, if the colony is small, it may be removed by slicing the wood
to which it is attached, the
slice to be placed in a watch-

Fig. 156.—Plumatélla.

glass of water on the microscope stage, which must, of
course, be in a horizontal position. The mirror may
then be swung above the stage, and *Plumatélla* viewed

by reflected light as an opaque object. It is exquisitely
beautiful in this position, as is Pectinatella or any of
the Polyzoa ; but the animals are very timid. To see
the expanded tentacles will therefore demand much
time and patience. Plumatella is almost as common as
Pectinatella. A board or log that has been floating
undisturbed in the pond will, during the summer, be
quite sure to afford a rich harvest of Plumatella if its
under-surface be examined.

4. FREDERICÉLLA.

The colonies of this Polyzoön are found in the
shadiest places and near the shores of shallow ponds,
growing like Plumatella, and often in company with it,
on the lower surfaces of floating or submerged objects.
The whole colony may be adherent, or only the base, the
stem and branches then floating. A single animal in-
habits each hollow branch, and resembles Plumatella in
appearance and structure. It may be distinguished
from Plumatella, however, by the oval or nearly circular lophophore, that of Plumatella being horseshoe-
shaped. The colonies are usually small, covering a small
space. The tentacles are never more than twenty-four
in number. The statoblasts are more or less circular,
and are without spines or hooks.

5. PALUDICÉLLA (Fig. 157).

These colonies may always be known from all other
tube-making Polyzoa by their jointed appearance, each

joint or cell being club-shaped. The colonies are irregularly branched, and are built up of a single row of cells placed end to end, the narrow end or the handle of the club being attached to the broad end of the cell immediately behind it. The opening through which the animal protrudes its circular lophophore is at one side

Fig. 157.—Paludicélla.

of the broad end of each cell near the top. The base alone may be attached, or the stem may be adherent and some of the branches free, as in the figure.

6. URNATÉLLA (Fig. 158).

The form and appearance of this Polyzoön are so characteristic that it need never be mistaken; but while the other members of the group are usually rather conspicuous objects to the eye of a microscopist, *Urnatélla* must be especially searched for. The colonies, or stem-like growths which it forms, are composed of urn-shaped cells or segments united end to end, and attached by a single disk-like enlargement to the supporting object from which they hang suspended. The lower surface of stones, beneath which the water constantly flows, seems to be Urnatella's favorite haunt. The stem-like colonies of urns are usually found two together pendent from the same disk of attachment, and appearing somewhat like a string of beads—this being due chiefly to the alternate bands of brownish-white and black surrounding the urns. In length the stems vary from one-eighth to one-sixth of an inch, rarely reaching one-fourth. To be seen on a wet stone with the unaided vision, therefore, demands a trained eye.

The cells or urns are joined end to end, the enlarged central portion of each being light-colored, while both the narrowed ends are dark or black. A single colony is seldom formed of more than a dozen urns, the stems thus built up being either quite straight or somewhat curved, or even on occasion loosely coiled. At times the stem is branched, the secondary stem originating near the point of attachment of one cell with the preceding, but soon falling off or voluntarily breaking away. On each side of every segment of the mature stems is a small, cup-shaped projection, the two appearing almost like handles to the urns. These are supposed to be the remains of the branches or of those segments which have fallen away and gone to begin new colonies in another place. Each urn, therefore, has at some time two urns attached to it, one on each side, and occasionally a specimen will be found with one or more branches still adherent.

The central enlarged portion of the urns is translucent, light-colored, and often with many transverse wrinkles and transverse brown lines. It is also brown spotted, and has many little tubercles of the same color. The necks of the urns where they are joined together to form the stems, are opaque and black. The first or foundation segment of the growth is larger than the others, and its base expands into a broad disk, which adheres to the stone and supports the entire stem. Through the centre of the whole collection of urns passes a cylindrical cord, whose purpose would seem to be to strengthen the fragile pile and to give it the great flexibility which it has.

The two segments near the free end of the stem are smaller than the others and rather different in shape. They are also nearly transparent and colorless. They seem to be urns in the process of growth, while those below are matured and hardened. It is only the terminal segments that contain the living animal, the urns forming the stem below them being filled with a soft, translucent, granular substance packed into the cavity around the central cord.

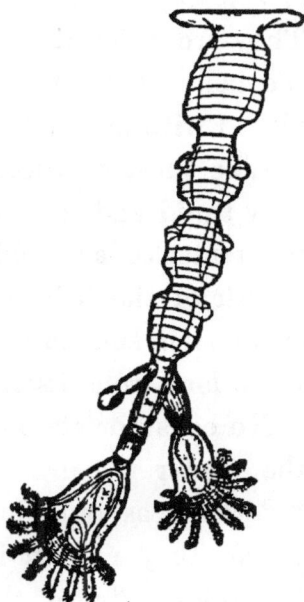

Fig. 158.—Urnatélla.

The animal that produces this beautiful series of brown urns lives at the free end of the pile solitary and alone with the exception of the temporary companionship of those short branches which sprout out near it, as shown in Fig. 158. It is these short growths that are supposed to drop off and leave the cup-shaped

11*

scars on each side. Rarely are there more than two of
these projecting scars on each urn. The animal itself,
which terminates the main stem and its branches, when
in active condition, appears, Dr. Leidy, its discoverer,
says, as a bell-shaped body with a widely expanded oval
or nearly circular mouth, directed obliquely to one side
or ventrally. The mouth of the bell is bordered by a
broad waving band or collar, from the inside of which
springs a circle of tentacles. Of these there are usually sixteen, though sometimes from twelve to fourteen.
They are cylindrical, and reflected from the mouth of
the bell. They are invested with an epithelium fur-
nished with moderately long, active cilia.*

Like most of these beautiful creatures, Urnatella is
very timid and sensitive. At the slightest disturbance
the tentacles are folded together and drawn into the
mouth of the bell, which closes around them, and the
entire stem suddenly bows itself down to the ground, or,
when long, rolls itself into a loose coil.

No eggs nor statoblasts have been observed. During
the winter the urns do not seem to become separated.
"Perhaps, as reproductive bodies, after the polyp-bells
perish, they remain in conjunction securely anchored
through the first of the series, and are preserved during
the cold of winter until, under the favorable condition
of spring, they put forth buds and branches, which, by

* "Urnatella gracilis: A Fresh-water Polyzoön." By Professor
Joseph Leidy. *Journal of the Academy of Natural Sciences of Philadelphia*, vol. ix.

separation and settlement elsewhere, become the foundation of new colonies."

Thus far I have not been fortunate enough to find a specimen of Urnatella, although it is probably not rare among stones in running water. A sight of its beautiful and curious collection of urns tipped by its graceful bell and swaying tentacles is worth many a long tramp, and careful scrutiny of many a wet stone. My account of this Polyzoön, because of my want of acquaintanceship with it, is gleaned from a paper, already referred to, by Dr. Leidy, who discovered and named it *Urnatélla grácilis.*

And those who desire to be fully informed as to the anatomy of the charming creatures which form the group of the fresh-water Polyzoa, and to distinguish the several species, are referred to Professor Alpheus Hyatt's work on "The Polyzoa," published by the Essex Institute, Salem, Mass., and to Professor Joseph Leidy's papers on the subject in the *Journal of the Academy of Natural Sciences of Philadelphia.*

CHAPTER X.

ENTOMÓSTRACA AND PHYLLÓPODA.

The reader is familiar with the crayfish, lobster, and crab as members of that great group of animals called the Crustacea, because they are covered by a hard, shelly coating; but, with the exception of the crayfish, he may associate them all with salt water, while in reality our fresh-water ponds are densely peopled with minute crustacean creatures. The little fresh-water animals are often enclosed in a bivalve shell, which some of them have the power to open and shut; or the back of the body may be simply hardened, but without a distinct shell. The feet, or legs, are usually numerous, and very hairy or bristly, and in one section of those referred to in this chapter are flattened, and each one bears near the body a flattened plate; consequently, since these parts are somewhat leaf-like, the animals have, as a class, been called the leaf-footed or the *Phyllópoda*, which is putting the words into Greek. Very many others, to be found much more abundantly and frequently than the Phyllopoda, are without these plates, although the feet are as numerous and, in some, almost as flat, and the shells or the shelly back as well marked. These have been, by naturalists, grouped together under the name of the *Entomóstraca*, meaning little animals in a shell, but the translation of the word has no distinctive signification, since members of both groups have shells.

The Entomostraca are more abundant in fresh water than the Phyllopoda, and are remarkably active. They are usually visible to the unaided eye as little specks, skipping, flirting, or jerking themselves through the water, although probably few will measure more than one-tenth of an inch in length. Under the microscope some are, as already stated, seen to be enclosed in a bivalve shell, and others are entirely free from so distinct a covering. The feet are arranged in pairs, and may be very numerous. They serve, in the shell-bearing forms, not only as swimming organs, but as gills or similar contrivances for the absorption of air from the water for the aëration of the little animals' blood. This is probably one reason why they are kept in such incessant motion. Even when the shell-bearing Entomostraca come to rest, to feed, or for some other purpose, certain of the feet keep up a ceaseless beating of the water, as can be readily seen through their transparent case.

The mouth parts are complicated, much patience and microscopical skill being needed to investigate and understand them. On each side of the head, however, and usually near the mouth, are two thread-like but jointed organs called the antennæ, and these the beginner must recognize, as they often become important aids in learning the animal's name. They vary in length, one on each side often being quite short and difficult to see distinctly, while the other two are usually long and conspicuous. They are all formed of short and well-marked joints, the number varying greatly in the different genera, and sometimes in different species of the same genus.

One or more black or dark red eye-spots are commonly present. In some the eye is single, and in the centre of the forehead. It may also be slightly movable at the will of its possessor. The young animal, as not rarely happens, may have two distinct eyes, which, as it grows older, become joined into one and covered by the shell.

The heart is very frequently visible, especially in the shell-bearing forms, being there placed at the back of the body and near the head. It beats rapidly, and apparently sends the colorless blood quickly through the system.

They all increase and multiply through the formation of eggs, which may remain within the shell and there be hatched, or they may be attached to the parents' body in external clusters. In the shell-bearing forms they are passed into a brood cavity at the back between the body and the shell, where they are kept until the young are hatched, when the latter make their escape into the water, and care for themselves. In those without shells the eggs are passed out of the body into one or two small, pear-shaped sacks called external ovaries, where they remain until hatched. In these cases, however, the egg masses are carried about by the parent, and are conspicuous objects. It is a common occurrence to find the little animals apparently loaded with the burden of eggs, and not uncommon to see the young escape. The "common Cyclops" is an instance. No member of the Entomostraca is so frequently seen and so abundant as the Cyclops, and hardly any other affords so good an example of this method of depositing and caring for the eggs in external ovaries, Cyclops having two of the latter, while some other almost equally common forms have but one. The external ovaries are usually long, pear-shaped bodies attached to the rear of the animal, near where it diminishes to form its tail-like portion. The eggs are round, unless they are made polygonal by pressure, almost black, and entirely opaque. In *Canthocamptus* there is but one external ovary. Both kinds are shown in Figs. 166 and 167.

The young, when first hatched, bear so slight a resemblance to the parent that some of them have been described and named as entirely different animals; and it was not until they were seen leaving the egg while still attached, or in the external ovary, that their true character was discovered. This is especially true of Cyclops. The young animal is shown in Figure 167*a*. It changes its skin several times before it begins to resemble its mother, a similar peculiarity being noticeable in many of the Entomostraca.

These little crustaceans are found in almost every body of still water. Some prefer the surface, where, on a sunshiny day, they are occasionally seen in immense numbers, sinking when a cloud shades them, and rising again to the sunlight. Others are to be taken only in deep water, while still others can be obtained only at night. Very many, however, are collected in every gathering of aquatic plants. They abound at all seasons of the year, even in midwinter. Their movements are rapid and characteristic. An Entomostracan can be readily recognized as such by the unaided sight, on account of the peculiar leaping, or short, jerking motions with which it travels through the water.

They are not only interesting little creatures to the microscopist, but they are extremely useful as well. They play a very important part in the food supply of fishes, forming the chief article of diet of some of our best fresh-water fishes. And they are almost as important as scavengers. Their favorite food is dead and decaying Algæ and animal matter, which, if allowed to remain in the great abundance in which it exists, our ponds and slow streams would before long become putrid and unbearable. But these numerous little creatures, by eating this refuse matter, transform it into an innocent and innocuous material, and confer a benefit both on themselves and us. Mr. C. L. Herrick, writing on this subject, says, " Their importance depends largely on their minute size and unparalleled numbers. The majority of non-carnivorous crustaceans are so constituted that their diet is nearly confined to such floating particles of matter as are present in the water in a state of more or less fine comminution; for, nearly without prehensile organs, these animals, by means of a valvular or, at most, ladle-like labrum, dip from the current of water kept flowing by the constant motion of the branchial feet, such fragments as the snail and scavenger-fish have disdained: bits of decaying Algæ, or the broken fragments of a disintegrated mosquito, all alike acceptable and unhesitatingly assimilated. The amount of such material that they will dispose of in a short period of time is truly astonishing."

When the shallow ponds are dried by the summer heat, the Entomostracans bury themselves in the mud, and there remain quiescent, but alive, so long as any moisture is present. When the mud is completely dried they die, but the eggs have the ability to endure heat and dryness without injury, and to develop and mature as the pools again become filled by the rain, or by the melting snow of early spring.

The Phyllopoda may also often be recognized without a microscopical examination, by their large size and almost universal habit of swimming on the back. *Branchipus*, sometimes called the fairy shrimp, and *Artémia*, or the brine shrimp, are nearly an inch in length.

As in the Entomostraca, their bodies may be incased in a bivalve shell or not. The broad, flattened feet are numerous, but the branchial or breathing-plates already referred to may be small and inconspicuous, and therefore difficult to be observed by the beginner. They are especially well-marked in Artemia (Fig. 169), and in

Branchipus (Fig. 170). Eyes are usually present, and large. In some forms they are elevated on stalks, thus reminding the observer of the stalked eyes of lobsters.

The eggs of the bivalve Phyllopoda are kept within a brood cavity, somewhat as in similarly incased Entomostraca, while in the shell-less forms they are carried about in a bottle-shaped sack at the end of the body, near the origin of the long, narrow, tail-like portion. In both kinds the young bear scarcely the remotest resemblance to the adults.

In the fresh and brackish waters of the eastern part of the country there are but few genera of the Phyllopoda represented, and none have yet been found in the ocean; while on the western plains and among the Rocky Mountains they abound. These latter forms are, however, not included in those referred to in the following list.

All these little crustaceans should be examined in a deep cell, to prevent the weight of the cover-glass from crushing their bodies. The shells and the shelly coating give them the appearance of hardness, but they are delicate and easily injured. The large Phyllopoda will need an especially deep and extensive cell.

The following Key will lead to the common genera of both divisions of these attractive animals. The only trouble the beginner may meet with in using it will probably be in determining whether the specimen is a Phyllopod or an Entomostracan; but as the former are large, and swim on the back, they may usually be deter-

mined by these appearances, and the name learned by
the Key, in connection with a pocket-lens. The two
Entomostracans, *Diáptomus*, and *Canthocámptus*, are
separated in the Key by the number of the joints in
their long antennæ. This seems to be a very minute
character to use in so artificial a table, but it need not
be an annoyance to the beginner, since the antennæ of
these two common little crustaceans differ so conspicuously in size and length that the joints need not be
actually counted; a glance will show which is Canthocamptus, with its short and rather inconspicuous antennæ.

The beak referred to is the front part of the shell
extended in a long, usually curved and pointed prolongation, containing the eye and portions of the animal's
head.

Key to Genera of Entomóstraca and Phyllópoda.

1. Legs with flat plates near the body; animal swimming on the back (*h*).
2. Legs without flat plates (*a*).
   a. Body enclosed in a bivalve shell (*b*).
   a. Body not enclosed in a shell (*g*).
   b. Shell with a sharp posterior spine or a tooth on
     or near the upper posterior angle (*c*).
   b. Shell without a posterior spine, or with one to four
     small teeth on the lower posterior angle (*d*).
   c. Smooth; spine on the upper angle, or near the
     middle of the border. *Dáphnia*, 1.

c. Smooth, brown ; spine on the lower angle. *Scapholéberis,* 2.

c. Reticulated ; spine on the lower angle ; antennæ large, cylindrical. *Bósmina,* 3.

c. Reticulated ; spine or tooth on the upper angle ; antennæ long, with two branches. *Ceriodáphnia,* 4.

d. Beaked in front (*e*).

d. Not beaked, oval, both ends rounded ; smooth or hairy. *Cýpris,* 5.

e. Posterior border with one to four small teeth. *Camptocércus,* 6.

e. Posterior border without teeth (*f*).

f. Shell nearly spherical ; posterior border truncate. *Chýdorus,* 7.

f. Shell not spherical ; posterior border convex ; antennæ small. *Alonópsis,* 8.

f. Shell not spherical ; posterior border truncate ; antennæ large, long, and branched. *Sida,* 9.

g. Body long and narrow ; antennæ long, twenty-five jointed. *Diáptomus,* 10.

g. Body long and narrow ; antennæ short, four to ten jointed. *Canthocámptus,* 11.

g. Body racket (battledoor) shaped, with two external ovaries. *Cýclops,* 12.

h. Body enclosed in a bivalve shell (*i*).

h. Body not enclosed in a shell (*j*).

i. Shell nearly spherical, smooth. *Limnétis,* 13.

i. Shell oval or oblong, flattened, amber-colored, with longitudinal lines. *Esthéria,* 14.

j. In brine pools and salt lakes ; eyes black, on stalks.
   Artémia, 15.

j. In fresh water ; males with large frontal appendages ; females without frontal appendages, but
   with an external, posterior, broad, short, and bottle-shaped egg-sack (*k*).

k. Frontal appendages much twisted and coiled ; body
   slender. *Chirocéphalus*, 16.

k. Frontal appendages not twisted nor coiled ; body
   stout. *Branchípus*, 17.

ENTOMÓSTRACA.

1. DÁPHNIA (Fig. 159).

There are several species of *Dáphnia*, all of which
may be known by the presence on the posterior border
of a sharp spine, which is never
on the lower angle. It varies in
length in the different species,
sometimes being nearly as long as
the shell, and extending obliquely upward. It also varies in length
and in position on the same individual, being longest in the young,
and becoming quite short with
age. In the species figured (*Dáphnia púlex*) it is usually on the

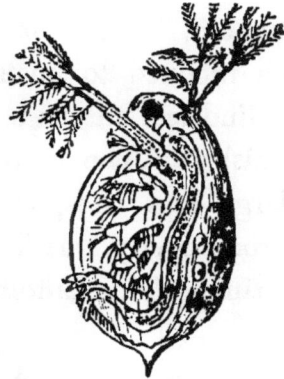

Fig. 159.—Dáphnia.

upper angle, but not rarely as shown in the cut. In
very old specimens it may be entirely absent, but it is

always present at some time of the animal's life. The
shell is oval and slightly flattened. The antennæ are
prominent, and are usually divided into two parts at
the free end, each division bearing several feathery
bristles. The feet are flattened, and generally in rapid
motion, so as to bring food to the mouth, and oxygen
to the blood. The heart is noticeable as a small colorless organ under the shell of the back near the head.
It pulsates rapidly. The eye is large and conspicuous.
The eggs are placed in a brood cavity, as shown in the
figure, and there hatched, the young being very different
in appearance from the parent. Daphnia is common
in the spring.

2. SCAPHOLÉBERIS.

The shell is somewhat beaked and usually dark brown.
The surface may be indistinctly reticulated or entirely
smooth. From *Bósmina*, for which the beginner may
be inclined to mistake it, the absence of the curved,
cylindrical antennæ common to that species will distinguish it. The posterior spines are short. The eye is
large and conspicuous. The egg is carried in the
brood cavity. It is said that but one egg is present at
a time. This Entomostracan is common.

3. BÓSMINA (Fig. 160).

The student will not have any trouble to recognize
Bósmina, on account of the long, large, cylindrical antennæ, each one curving downward from the side of

the head like the trunk of a microscopic elephant. The shell is oval, colorless, and the posterior border has a spine at its lower angle, never at any other point. The net-work of lines on the surface may extend over the entire shell or be restricted to some one part. The eye is large. The eggs are hatched

Fig. 100.—Bósmina.

in a brood cavity on the back beneath the shell. The heart is visible near the centre of the back. Bosmina is not so common as Daphnia.

4. Ceriodáphnia.

The shell is oval, oblong, or somewhat four-sided, and always beautifully, if coarsely and conspicuously, reticulated, the meshes being hexagonal and comparatively large. The head is separated from the body by a depression in the shell, and just behind the rather small eye-like spot it has a slight elevation. The eye is usually near the rounded lower margin or tip of the beak-like head. The antennæ resemble those of Daphnia, being long, and divided into two three-jointed branches of equal length. The angle or tooth on the upper corner of the posterior border is usually sharp and conspicuous.

This Entomostracan is abundant in the writer's locality. It is visible to the naked eye, being about one twenty-fifth of an inch long. In the aquarium its movements are almost distinctive. It seems to prefer the centre of the vessel, where it darts upward for a

short distance with a jerk, only to allow itself to float
back to the starting-point. A glass jar well stocked
with these pretty creatures leaping up and down irregularly and incessantly is an interesting sight. Under the one-inch objective the little animal is more than
interesting.

5. Cýpris (Fig. 161).

The shell entirely surrounds the animal, so that the
little creature, when danger threatens, shuts itself in as
completely as a clam or a mussel, and allows itself to fall
to the bottom. The form varies from an oval to a kid-
ney shape, according to the species, and the color may
be green or brown, or whitish and
marked with several dusky bands. It
may be smooth, or entirely covered
with fine hairs, or only the free borders may be fringed. The shell is

Fig. 161.—Cýpris.

never opened wide, but the legs and feathery antennæ
project from a narrow cleft between the valves, the little animal swimming rapidly by their aid, or creeping
about the slide or over the aquatic vegetation. *Cýpris*
is reproduced by eggs, but "the mass of eggs, including
about twenty-four, is attached by the female to water-
plants with the aid of a glutinous secretion, an operation which lasts about twelve hours."

6. Camptocércus (Fig. 162).

The shell is elongated, somewhat quadrangular, transparent, and marked by lines traversing the surface

lengthwise. The beak is blunt, and usually curved downward, or it may extend slightly away from the body. The head is strongly arched. The teeth on the posterior border (not shown in the figure) are small, and vary from one to four. The eye is small. The eggs are carried in a brood cavity. The animal occurs chiefly in lakes and large ponds.

Fig. 162.
Camptocércus.

7. Chýdorus (Fig. 163).

The surface of this nearly spherical shell is usually reticulated. The beak is long, curved, and pointed, being sharp in the female. The posterior border is truncate in young specimens, becoming more rounded in the old. The eye is present and single. The eggs are hatched in the brood cavity. The animal occurs abundantly very early in the spring, usually near the bottom, living chiefly on vegetable matters. The motion is rolling and somewhat unsteady, and uncertain in appearance.

Fig. 163.—Chýdorus.

8. Alonópsis (Fig. 164).

The lower or free edge of the shell is fringed with bristles, which are longest in front. The beak is long, pointed, and separated by some distance from the body of the shell. Eye large. One of the feet (the third) has a long spine fringed with

Fig. 164.—Alonópsis.

12

short hairs on the edges, and often reaching to the posterior margin of the shell. The surface is usually marked by a few conspicuous diagonal lines. The animal's movements are slow.

9. SÍDA.

The shell is long and narrow, with the head separated from the body by a depression. The posterior margin is nearly straight, and has no spine or tooth. The antennæ are large, and somewhat resemble those of Daphnia, although in *Sida* they are rather stouter, and are divided into two *unequal* branches. There is but one species — *Sida crystallina*. It is quite common in some localities.

10. DIÁPTOMUS (Fig. 165).

Diáptomus may be recognized by the very long antennæ, which are often as long as the body. The latter, including the head, is formed of six joints, and the posterior narrower part or abdomen of five, although in the female two of the latter may be united, thus giving it a three-jointed appearance. The animal is among the largest of the Entomostraca, often measuring one-tenth of

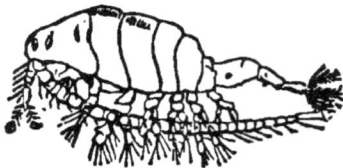

Fig. 165.—Diáptomus.

an inch in length. The color is often brilliant, varying in the different species, and even in the different parts of the .body of the same individual. It may be deep red, brilliant purple, bluish with purple-tipped antennæ, whitish, or colorless. The animals may be found in shallow pools in the fall and early spring, and occasionally in slowly flowing streams. The external ovary is single.

11. Canthocámptus (Fig. 166).

After Cyclops and Daphnia this is the commonest fresh-water Entomostracan in the writer's vicinity. A gathering of aquatic plants can seldom be made in this neighborhood without obtaining many of the graceful little *Canthocámpti.* They are visible to the unaided eye as small, flesh-colored, or pinkish lines darting through the water in short jerks, after the manner of most Entomostraca. Like all minute animals, they will collect on the best lighted side of the bottle, where they may be easily captured with the dipping-tube. The eye is single. The antennæ are short and quite hairy. The body is long, narrow, and subcylindrical, being widest and thickest in front. There is no distinct heart. The external ovary is single. It is attached

Fig. 166.—Canthocámptus.

to the parent by the thinnest and apparently most delicate part, although considerable force is necessary to separate it from the body. The eggs are round and opaque. The young differ greatly from their mature aspect. Canthocamptus is found in almost any shallow body of still water, and all the year through, even occasionally in midwinter. It is shown in side view in the figure, so as to exhibit the single external ovary so characteristic of it.

12. CÝCLOPS (Figs. 167, 167a).

This commonest of all fresh-water Entomostraca has a single eye in the middle of the forehead, like the giants of ancient story, a bifid tail adapted for swim-

Fig. 167.—Cÿclops.

ming, and two external ovaries, one on each side. These ovaries are long, pear-shaped sacks filled with dark, opaque eggs, and at-tached to the body by the narrow or stem end of the pear. The young (Fig. 167a) pass through several stages before they begin to resemble the parent. It has been said that the eggs are carried in the external ovaries only until they are ready to hatch, when they are deposited before the young make their escape. This is a mistake, as the student will probably soon observe. The young leave the eggs while they are still attached to the parent. They break the egg membrane very suddenly and unex-pectedly, although the observer may have

Fig. 167a.
Young Cÿclops.

been for some time watching the little creatures rest-lessly moving about inside. As they escape they often dart half-way across the field of a low-power objective.

If *Cÿclops* had no enemies the waters would soon become filled with them in numbers almost beyond imagining. One female Cyclops has been seen to lay ten times in succession; but, to be within bounds, the observer who made the calculation supposes a single one to lay eight times only, and forty eggs at each time. "At the end of one year this female would have been the progenitor of 4,442,189,120 young — that is, near four and a half thousands of millions."

There are about thirty species of Cyclops, and in all of them there are four antennæ, two being long and conspicuous, the other two small, and often carried so that they are invisible unless the Cyclops is turned on its back.

PHYLLÓPODA.

13. LIMNÉTIS (Fig. 168).

The oval or nearly spherical, smooth shell has a well-marked beak, which in some of the species is enormous, while in others it is less conspicuous. When the valves are closed they measure about one-sixth of an inch in length, and have often been mistaken for small fresh-water mollusks of the genus *Pi-sidium.* The eyes are two, but so

Fig. 168.—Limnétis.

close together that they often appear to be united; they are black. The animals swim on the back, as do so many of the Phyllopoda. In the males the two front legs are flattened, and have on the end of each a complicated organ called the hand, although it bears the most remote resemblance to the human hand. The eggs are carried on the back under the shell. The animals are flesh color.

14. ESTHÉRIA.

The shell is smooth and shining, and marked with distinct lines running almost parallel with the front, or free edge, of the valves. It is very thin, flat, and large, measuring about two-thirds of an inch in length. The males have two pairs of hands, or one on each of the four front legs. The shell of the several species varies from oval to oblong with the upper margin very much flattened, or it may be somewhat globose. Most of the species are confined to the waters west of the Mississippi River, one, however (*Esthéria Mexicána*), being found near Cincinnati. Many of them are in appearance not unlike a small clam, or the little fresh-water mollusk, Pisidium, so common almost everywhere.

15 ARTÉMIA (Fig. 169).

Fig. 169.—Artémia (a female).

Artémia occurs only in brine or the water of salt lakes. It is not rarely found in the hogsheads of water on railroad bridges or trestles, where the water is made salt to prevent freezing. The bodies are slender and pale red, flesh-color, or sometimes greenish. The feet are eleven pairs, beautifully fringed with many long hairs, and bearing the flattened branchial or breathing-plates. When the creature swims on its back, as it habitually does, these feathery feet beat the water in rapid succession, as if a wave of motion were rapidly passing above them. It is a beautiful creature, and one sure to attract attention, not only by its graceful motions and preference for salt water, but by its size, being half an inch or more in length. The eyes are black, and placed on the ends of stalks projecting from each side of the rather small head. The antennæ are short, but conspicuous. The eggs are yellowish-white. The young are very active, and differ much in appearance from the parent. They are blood red, with one bluish eye.

16. Chirocéphalus.

This curious creature has eleven pairs of swimming feet, as has Branchipus, but there need be no difficulty in distinguishing it from Branchipus (for which it may be mistaken) provided the male is obtained. If the female alone is captured some trouble may be experienced by the beginner in determining one from the other. The female of *Chirocéphalus*, however, is slender, while that of Branchipus is stout; but such a distinction is valueless until both have been seen, or the two sexes have been taken from the same pond. In the latter case the male may be known by the two remarkable appendages hanging down from the sides of the head. These are about one-fourth of an inch long when extended, and are curved and coiled and twisted in a way that defies description. Each one is broad near the upper or attached ends, and diminishes to a long, curved point covered with minute spines, while in its entire

length it is curiously lobed. The egg-sack of the female is short and small, and the attached end is lengthened, somewhat like the neck of a bottle. The eggs are
very large, and about twelve in number. The body of
each sex is about two-thirds of an inch long. Chiro-
cephalus is often found in company with Branchipus,
usually in the spring, as early as the middle of March.

17. BRANCHÍPUS (Fig. 170).

The flesh-colored or pale red body is stout and large,
often measuring an inch in length. The head is large,
and the frontal appendages of the male are long and
broad, as shown enlarged in Fig. 170. These hang down
on each side of the head,
and are formed of two quite
dissimilar parts. The upper
half is broad and thick, and
about one-fifth of an inch
long. It ends in a stiff,

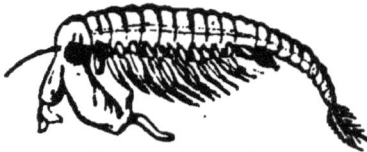

Fig. 170.—Branchipus
(a male).

bristle-like prolongation of nearly equal length, with a
short, bristle-like tooth at the inner side at the point of
junction of the two parts. There are eleven pairs of
swimming feet, and the animal swims on the back. The
eyes are two, black, and elevated on the ends of short
stalks. The body of the female is as large and as stout
as that of the male. The egg-sack is noticeable near the
point of union between the posterior narrow portion of
the body and the broader front.

It is a curious fact that *Branchipus* is killed by even

the heat of early summer or late spring. Dr. Packard, describing a visit to a pond where these creatures had been found on May 2d, but from which they had all disappeared by May 13th, says, "It seems from this quite evident that the animal probably dies off at the approach of warm weather, and does not reappear until after cool weather sets in late in the autumn, being represented in the summer by the eggs alone; and thus the appearance and disappearance of this Phyllopod is apparently determined mainly by the temperature."

A vessel full of water in which Branchipus is floating on its back is a strangely beautiful and interesting sight. The pale reddish or flesh-colored bodies rising and falling in long curves, with their numerous broad feet waving together rhythmically, make a living picture long to be pleasantly remembered.

Those readers who desire to pursue the subject, especially in respect to the anatomy and development of these crustaceans, are referred to Mr. C. L. Herrick's "Crustacea of Minnesota," published in the twelfth annual report of the State Geologist, and to Prof. A. S. Packard's "Monograph of the Phyllopod Crustacea of North America," issued in the twelfth annual report of the United States Geological and Geographical Survey of the Territories, Dr. F. V. Hayden in charge.

12*.

CHAPTER XI.

WATER-MITES AND THE WATER-BEAR.

THE Water-mites (Fig. 171) are sometimes called water-spiders, probably because they bear some resemblance to small spiders, and have eight legs. Naturalists have seen the resemblance and have placed them in a family group near to the spiders. Water-spider, however, is not a good name for them, as we have some true spiders that are semi-aquatic in their habits and have therefore a better title to such a name.

The water-mites are usually very active little animals, swimming freely and rapidly through the water, or forcing themselves among the leaflets of aquatic plants, probably in search of food. They may generally be obtained in some abundance by collecting water-weeds in the way previously recommended, namely, by sinking the bottle and floating the plants into it without removing them from their native element. They are

Fig. 171.—A Water-mite.

all quite visible to the unaided eye, and may for the most part be studied with a comparatively low-power objective.

Their bodies are plump and oval, or nearly spherical.

The skin of most of the forms is soft and easily broken, but in the members of a single genus, *Arrenúrus*, the surface is firm and comparatively hard. They are all brightly, even brilliantly, colored. They may be of one uniform tint, with a few blackish or brownish spots on the posterior region, or the single individual may be variously tinged in different parts of the body. The colors are of almost every imaginable shade of crimson, azure blue, yellow, green, brown, gray, or purple. The eight long legs also share in the general brilliancy, and often present a coloration entirely different from that of the body.

The eyes are usually on the upper surface near the front border. They are small, and may be either round or crescentic in shape, red, black, or carmine in color, and two or four in number. They are usually placed close together, and when four in number, are arranged in two distinct pairs.

The upper part or the back of the little animals may be entirely smooth, densely clothed with short hairs, or with a few scattered, fine bristles. It may also present no markings when magnified, or, as in a single genus, Arrenurus, it may be beautifully ornamented with a net-work of narrow meshes in a hexagonal pattern. In all, or nearly all, of the mites the upper surface bears two or more black, dark brown, or reddish spots quite distinct from the general coloring of the body. These are caused by nearness to the surface of the intestine or other internal viscera, the dark contents of which show

their color through the skin.   In some these dark spots
become large, occupying much of the upper surface, and
so arranged and shaped that they leave between them
in the middle line of the body a ϒ-shaped space which
may be white, yellow, or other color.   These spots are
called cœca or the cœcal markings, the word being the
plural of cœcum, meaning a certain part of the intestinal
canal.   They are useful to the student in identifying
the species.

The lower or ventral surface is the most important
part to the observer who desires to ascertain the name
of his specimen, or to the student who wishes to make
a more serious study of the animals, for on this surface
are the parts most used by the naturalist in classifying
the mites.   The beginner must therefore seek to have
the little creature arranged on its back before it is
placed under the microscope, so that the ventral surface
shall be presented to the objective.   This is sometimes
a difficult operation to accomplish without injuring the
delicate body.   The writer has used for the purpose a
little home-made contrivance that answers well and can
be made by any one.   A hole about half an inch in diameter is drilled through a glass slip, and into one side
is cemented with shellac a thin glass circular cover a
little smaller than the hole, so that the thin cover may
not be flush with the surface of the slip.   It is not
very difficult to grind a hole through a thin glass slip
if the file or other grinding tool is kept wet with tur-
pentine.   The aperture may not be a perfect circle; it

will probably be very irregular—I know mine is—but it will answer every purpose. The mite is placed in this cell, and a thin cover applied to the opposite side, thus forming a glass box that can be turned over for the examination of both surfaces of the animal, and is deep enough not to injure the soft body, yet shallow enough to restrain its movements.

The mouth of the mite is usually a complicated affair, and is sometimes surrounded by a circular elevation or ring called a hood, and always having short, jointed palpi, or feelers. At some distance back of the mouth, in some forms quite near to the posterior border, but always in the median line, will be seen in the female mites a small dark spot or narrow line which is really an opening. In some this orifice, which may be called the ventral opening, is covered and concealed by a large plate, called the ventral plate; or there may be two plates, curved, oval, or other shape, one on each side of the ventral opening. They are useful to the naturalist as one means by which the mites may be classified, and they should be carefully searched for by the beginner who desires to learn the name of his specimen. They are not present in the males. The reader will therefore perceive that to identify his captive the specimen must be a female. The two sexes, however, differ so conspicuously in appearance that they are easily recognized. The female always has the posterior border of the body more or less evenly rounded, while the male frequently possesses a peculiar little tail-like process projecting from the middle of the rear margin. One form of this curious projection is shown in Fig. 176, a female by Fig. 171, the projection varying in shape and size in the different species. The males seem much less abundant than the females; they are, at least, less frequently captured by the microscopical fisherman.

On the ventral surface, behind or before, or on both sides of the ventral plates, will be observed one or more very small dark spots never bordered by a plate. These are the external openings of the tracheæ or air-tubes, which extend through the body and supply it with oxygen. As the mites are not known to come to the surface for a supply of air, as so many aquatic animals do, the tracheæ are supposed to be able to absorb it directly from the water. The tracheal openings are not an important aid in ascertaining the name of the creature, but the beginner must not mistake them for the aperture bordered or covered by the ventral plates. In some mites they are not well marked, and may be overlooked, and there is still another dark spot usually present near the posterior part of the ventral surface which must not be confounded with the ventral opening since it is in the median line. This is the external opening of the intestine. It is never bordered by plates, and is always behind the ventral orifice, but it is not always conspicuous.

Equally important to the student are certain elevations of the ventral surface which appear to cover the attached ends of the legs. These are called the *coxæ*, the plural of *coxa*, a Latin word meaning the thigh. They are variously shaped and arranged, one coxa seeming to cover the end of each leg, or appearing to be the thigh belonging to that leg. They are motionless, however, and are really only elevations of the skin, beneath which the muscles of the legs may be seen in action. In some mites the coxæ on each side of the body are arranged in groups of two each, the borders of the two which form the group being in contact either by their whole length, as in Figs. 175, 176, and 177, or only at some single point, as in the posterior group shown in Fig. 174. In Figs. 174, 175, 176, and 177 there are four . groups, formed of two coxæ each; in Fig. 173 there are six groups, the anterior alone being formed of two, the two posterior groups on each side being of but one coxa each and separated. Their shape differs widely even in the species of one genus; their arrangement, however, is constant and important.

The eight legs are long and jointed, the last joint ending in one or two short claws. The hairs fringing their margins are long and numerous, and are used as aids in swimming. They add a good deal to the beauty of the animal.

Mites are found in salt as well as in fresh water, but with the marine forms this little book has nothing to do. The fresh-water ones are propagated by means of eggs, which are often seen attached to the stems of aquatic plants or to the lower surface of floating leaves, where the writer has obtained them and had them to hatch in captivity. They are small, brownish, jelly masses, which might easily be overlooked or passed by as snails' eggs, often to be found in the same localities. The newly-hatched young often bear but a slight resemblance to the parents, those of some genera having but six legs, those of one species being said to have but three. Many of these immature forms are parasitic on aquatic insects, becoming free-swimming and independent when they attain adult growth and age. Some of the mature mites are also parasitic in the gills of the fresh-water mussel (*Unio*). On account of these peculiarities the study of their life history is a difficult one.

The Entomostraca and Infusoria are said to form their favorite food.

There may seem to be but little connection between the water-mites and the water-bear, and still less resemblance, yet naturalists have classified them near together. The water-bear (Fig. 172) is a common and curious aquatic animal, so closely and so comically resembling a transparent eight-legged microscopic bear that the beginner will know it the first time he sees it; further reference to it is therefore reserved for another page (p. 267).

Key to Genera of the Water-mites (Hydrachnidæ).

1. Body colorless, cylindrical, elongated, and transparent; legs eight, short, with claws; the animal walks slowly and is bear-like in appearance. *Water-bear (Macrobiótus)*, 1.

2. Body brightly colored, oval, or spherical; legs eight, long; animal swimming actively (*a*).

3. Body brightly colored, oval, or spherical; legs eight, long; animal walking, never swimming (*e*).

   a. Ventral plate single, cordate, the apex pointing forward. *Diplodóntus*, 2.

   a. Ventral plate single, cordate, the apex rounded, pointing backward (*b*).

   a. Ventral plate double (*c*).

   b. Posterior coxæ on the same side not in contact. *Hydráchna*, 3.

   c. Posterior coxæ on the same side in contact by their whole length (*d*).

   c. Posterior coxæ on the same side in contact only by their internal ends, their outer extremities diverging. *Eyläis*, 4.

   d. Ventral plates oval, with an oval plate on each side; mouth round, with a circular hood. *Arrenúrus*, 5.

   d. Ventral plates narrow, curved, each with two or three translucent tubercles. *Átax*, 6.

   e. Eyes four, on a lanceolate plate; coxæ in four groups. *Limnóchares*, 7.

1. THE WATER-BEAR: *Macrobíotus* (Fig. 172).

The body is soft, colorless, and transparent. The legs are very short, and have on the end of each several sharp claws, the legs being arranged three on each side of the body and two at or near the posterior extremity. The

mouth is a small opening at the front of the part repre-
senting the head. It is followed internally by two short,
somewhat curved and diverging rods, said to be used to
wound the prey. The so-called gizzard, at a short distance from the mouth, is plainly visible through the
transparent body. It has no motion. Two small eyes
are usually present, one on each side of the head. The
animal's movements are very slow and awkward, the creature appearing to work hard,
with but little result so far as progress is
concerned.

Fig. 172.—The
Water-bear
(Macrobiótus).

Macrobiótus is produced by eggs, which
are deposited in an interesting way. When
they are sufficiently matured, the water-bear
sheds its skin and leaves the eggs in the empty and cast-
off case. It is no unusual occurrence to find the empty
skin of Macrobiotus with the empty eggs inside, the
young having escaped. The young resemble the parent, it is said, in all except size.

This strange, bear-like creature is to be found quite
often at the bottom of shallow ponds; or, if an aquarium is kept, it will be almost sure to make the bottom
its home. It is entirely invisible to the naked eye, measuring rather less than one-sixtieth of an inch in length.
On account of their slow movements, the water-bears
are often called *Tardigrades*. The scientific name of
the common American form is *Macrobiótus Ameri-
cánus*.

2. DIPLODÓNTUS.

This mite may be recognized by the form of the ventral plate as given in the Key, and by the fact that the plate is roughened by minute granules. The eyes in one species are two in number, very small, and wide apart. They are placed on the edge of the front border. In another species they are four, and are placed so far forward on the front margin that they are best seen when the animal is on its back, and thus examined from beneath. The coxæ are in four separate groups. The body of the two-eyed species has the front part black, spotted with red, and the posterior half red, with a central longitudinal black band. The one with four eyes has the body bright red.

3. HYDRÁCHNA (Fig. 173).

The anterior coxæ on the same side form a single group, being in contact by their whole length; the middle one is entirely disconnected from the others; the most posterior is the largest, and is also entirely separate. In one species the body is spherical and black, with yellow dots, the legs being shorter than the body, and black, with red ends.

Fig. 173.—Coxæ of Hydráchna.

In another the body is red, with two pairs of dark red eyes, and long legs. The young are said to have but three legs.

4. EYLÄIS (Fig. 174).

The two anterior coxæ are in contact by their entire length, and form one group on each side. The two posterior coxæ are in contact only as described in the Key. They are all moderately narrow. The mouth is round, ciliated, and with a kind of hood which the beginner may have some trouble to recognize. The ventral plates are curved, almost crescentic, and narrow, one being on each side of the ventral opening, and just behind them are two small tracheal apertures. The intestinal orifice is visible at the rear in the middle line, with a tracheal opening on each

Fig. 174.—Coxæ of Eyläis.

side. The eyes are four, in two pairs, rather close together. A large red, nearly spherical *Eyläis* is quite common in our ponds. The young are described as being red, transparent, with four eyes wide apart, and six legs.

5. ARRENÚRUS (Figs. 175, 176).

The coxæ form two groups on each side, the two anterior coxæ being in contact by their entire length, as are also the two posterior. Occasionally the anterior groups on opposite sides are in contact at the median line, as in Fig. 175. The ventral plates are oval, their greatest length generally being from before backward, and usually close together. The two oval, lateral plates are oval from side to side, and sometimes curved. The mouth is small, round, and encircled by a ring-like hood. The skin is usually hard and roughened, or covered by a deep net-work of strongly elevated lines, which give it a beautiful

Fig. 175.—Coxæ of Arrenúrus (female).

appearance. The skin is sometimes shed in captivity, and is not rarely found as a torn, empty, and colorless net, quite worth examining with a high power. There are some soft-bodied species, but they do not seem to be common.

The body of both the male and female is truncated at the posterior border, but the male has a peculiar short prolongation projecting from the centre of that margin, as in Fig. 176, the shape of the part varying greatly in the different species. The females are the most numerous and most frequently met with.

Fig. 176.
Coxæ of Arrenurus (male).

The upper surface or back of both sexes often bears a deep, depressed line, sometimes enclosing a small circular or oval area confined to the posterior extremity, sometimes a large space, including the greater part of the entire back. From others it may be absent. The eyes are two, black, and separated. The color of the body is very different in the numerous species. It may be blue, green, yellow, red, or almost any bright tint, either diffused or confined to distinct parts. Thus in one female the centre of the body is brown, the sides blue, and the coxæ yellow. In another the body is red, the cœca vermilion. Another, a male, has a bright yellow tail, the centre of the body white, with a blue line near the posterior border, and dark brown cœca.

A species of Arrenurus with a hard and reticulated surface is not uncommonly captured among Ceratophyllum and Myriophyllum in rather shady places which,

I think, all the water-mites prefer. A moment's exposure to the direct sunlight out of the water is fatal.

6. ÁTAX (Fig. 177).

The anterior coxæ are in contact by their entire length, and the posterior extremities of the two anterior groups on each side are also often in contact, thus appearing to press the mouth between them. The posterior coxæ are also in contact for their whole length, but they do not touch those of the opposite side of the body. The

Fig. 177.—Coxæ of Átax.

fourth coxa, or the one belonging to the most posterior leg, is usually much larger and broader than any of the others. The two ventral plates are narrow and curved, the tubercles on each of them being rounded and translucent. The front pair of legs are long and curved, the hairs on them being bristle-like. When the mite walks these legs are held rigidly in front. The color of the body varies, as it does in the other forms. A yellow *Átax* is not uncommon in our ponds and shallow, slowly flowing streams.

7. LIMNÓCHARES (Fig. 178).

This mite may be recognized by its habit of always walking. It never swims. In this it differs from all the other known forms of fresh-water mites. The eyes are four, and are arranged on a lanceolate plate (Fig. 178, much enlarged), two on each side, with a central rounded projection between them in front. They are also surrounded by hairs. The coxæ do not make the prominent elevations common to the other mites, but seem rather to be beneath the skin. "The coxæ of the anterior two pairs of legs are closely approximate, as are also those of the two posterior pairs, but the two groups are widely separated." The anterior are larger than the posterior. The mites are

Fig. 178.—Eye-plate of Lim-nóchares.

small. I have never been so fortunate as to meet with *Limnóchares* nor with *Diplodóntus*.

If the observer should desire to make permanently mounted slides of his specimens of mites, he may try a preservative medium prepared by mixing eight parts of water containing a drop or two of carbolic acid, with one part of glycerine. This is said to keep the bodies without the loss of their characteristic plumpness, nor much of their color, if mounted in a deep cell. Two specimens should, if possible, be preserved in the same cell, and so arranged as to show both surfaces. They will usually need a large and deep ring.

The fresh-water mites have never been studied by any American naturalist; there are, therefore, no books nor even isolated papers to which the beginner may be referred for further aid. The field is an entirely unexplored one so far as the water-mites of this country are concerned. To a student with an eye sensitive to color, and with a large amount of patience, the subject ought to be an attractive one.

CHAPTER XII.

SOME COMMON OBJECTS WORTH EXAMINING.

THERE is literally no end to the objects worth examining with the microscope. Even a pocket-lens reveals new and wondrous aspects in the most familiar things. An old and leathery lichen from a stump becomes a charming picture and a living one, for its hills and hollows and winding valleys are the homes and hiding-places of innumerable little creatures which the pocket-lens brings to view. The furrowed and weather-worn bark of any tree has countless points of interest and charm. Neither need there be any scarcity of material at any season of the year. In the spring and summer the only trouble is to find the time to examine even a small portion of all the wondrous and beautiful things that nature then offers. In midwinter, life and beauty are almost as abundant, if less conspicuous. The dust swept by a feather from the wall of a dark cellar may bring to light minute creatures not to be obtained elsewhere. There is no end to the objects, there is no end to the unlikely places that will reward a little search, and there is no end to the telling, unless it be an abrupt one.

But although the beginner may never be at a loss for

employment for his pocket-lens, he may at first feel that
his compound microscope will never afford him either
amusement or instruction. How glaring and how laugh-
able will such a mistake appear after six months' use
of the instrument! When recommending a friend to
purchase a microscope, he will speak of that conclusion
as an amusing episode in his life. Yet the beginner,
especially if alone, or if without even a friend to sug-
gest, or an experienced microscopist to instruct, must
necessarily be somewhat at a loss as to how to make a
start, and I know of no remedy for this unpleasant feeling except to experiment. Take the first small object
that may be convenient, place it on a glass slip, and examine it with a low-power objective; add a drop of water, cover it with a thin glass square, and note the change
in its appearance. But do not imitate the man who returned his microscope to the manufacturer because, as
he said, it would not show the crystals in sugar. After
much questioning it was discovered that he had placed a
huge lump of loaf-sugar on the stage, expecting that the
crystals would at once be conspicuous. And do not imitate the man who put on his stage a piece of anthracite
coal direct from the bin, expecting it to reveal its vegetable nature and fern fragments or impressions without
any previous preparation. A lump of sugar or of coal
is a dark object when an attempt is made to throw light
through it by a microscope mirror, yet both are beautiful and interesting when viewed as opaque objects,
with the light reflected on them from the mirror swung

13

above the stage.   To see the sugar crystals, however, or
the structure of coal, demands some careful and skilful preliminary preparation of the materials.   The coal
must be sliced and ground down until it becomes transparent, or at least translucent, an operation that needs
an expert to accomplish.   Sugar crystals can be seen by
any one, as the reader will presently learn.

The smaller the object the better it usually is for microscopic examination.   The field of even a low-power
objective is much smaller, and consequently includes
less of the object than the beginner would imagine.
Try the experiment by placing a piece of hair a quarter
of an inch long under the one-inch objective.   You will
not be able to see both ends of the hair in the field at
the same time.   The stage must be moved for what appears to be a long distance before the end is brought
into view.   With the one-fifth objective the field is still
smaller, and the higher the power the smaller the field.

The object should also be as thin as possible, so that
the light may pass through it, unless it is to be viewed
as an opaque object, with the light thrown on it from
the mirror above the stage.   In that case the thickness
makes but little difference, if the beginner will remember that the microscope is for the study of small things.
In opaque objects, however, only the surface can be examined—an important and often a beautiful part—but
in transparent substances the internal structure can be
seen.

The following is a short list of common objects of interest to the beginner, and, indeed, to any one. They are all very accessible in the writer's locality; it is hoped that they are equally accessible to the reader. The list is intended only as a start, or a hint as to what may be examined, and be interesting yet common and abundant. The beginner will soon cease to be a beginner. He will before long become so interested in some special class of nature's handiwork that he will leave all the others and devote himself to that one. No single student can expect or hope to cultivate all departments of microscopical science. The field is too vast and life is too short. Most heartily would I recommend the beginner in the use of the microscope to spend several years, if necessary, in taking short excursions into as many different microscopical departments as possible, and to then intelligently make a selection of some one scientific field, of which there are many, and make its cultivation the work and the recreation of his leisure hours. The work will soon become recreation, and the recreation will soon result in increased knowledge not only to the student-worker but to the scientific world at large. There can be no better way of employing one's leisure hours than by scientific work or even by scientific play. The illustrious Leidy, one of the greatest of living naturalists and investigators, says in his monograph on the fresh-water Rhizopods of North America, "The study of natural history in the leisure of my life, since I was fourteen years of age, has been to me a constant source of happiness, and my experience

of it is such that, independently of its higher merits, I
warmly recommend it as a pastime, than which, I be-
lieve, no other can excel it. At the same time, in ob-
serving the modes of life of those around me, it has
been a matter of increasing regret that so few, so very
few, people give attention to intellectual pursuits of any
kind. In the incessant and necessary struggle for bread,
we repeatedly hear the expression that 'man shall not
live by bread alone,' and yet it remains unappreciated
by the mass of even so-called enlightened humanity.
In common with all other animals, the engrossing care
of man is food for the stomach, while intellectual food
remains unknown, is disregarded or rejected."

1. SCALES FROM INSECTS' WINGS.—Butterflies' wings
are profusely covered by minute scales of great beauty
of form and color. They differ widely in different
kinds of butterflies, and often on different parts of the
same wing. Gnats and mosquitoes are also a source of
supply for still more minute and curious scales. To
obtain them from the butterfly, brush the wing with a
small camel's-hair pencil, or gently scrape it with the
point of a penknife, and examine the fine dust that
falls. Gnats and mosquitoes may be allowed to fly
about in a small, perfectly dry phial, and the scales thus
knocked off are to be transferred to the slip by invert-
ing the bottle and tapping it against the glass. The
more mosquitoes imprisoned, of course the more numerous will be the scales. The common clothes-moth is

also a scale-bearer. The abundant *Lepisma saccharina*, a little, flattened, fish-shaped, silvery creature, often seen running about old books, is covered with scales which were at one time used as test objects. They may be obtained by tapping the animal against a slide, or by scraping the surface gently. All scales may be examined and mounted dry. A small piece cut from the wing of a butterfly and viewed as an opaque object, will show the arrangement of the scale-like feathers, a sight well worth seeing.

2. ÉLYTRA.—The hard wing-covers (*élytra*) of beetles are often of startling beauty when examined with a strong light as opaque objects. They are not transparent, and cannot be made so, consequently only the surface can be viewed. They are to be examined and mounted dry.

3. INSECTS' FEET vary greatly in appearance and structure. They afford an endless supply of the most interesting objects. They should be cut off with sharp scissors, so as to include part of the leg, and examined in water. If the air clings to the bristles on the leg, dip the latter into a drop of alcohol and transfer it to the water on the slide before the alcohol evaporates. The feet of the two or three kinds of flies that frequent our houses, the feet of spiders and those of caterpillars, should not be overlooked by the beginner. Spiders can be obtained at all seasons, even in the winter, by searching the dark corners of the cellar where the furnace is.

4. EYES OF INSECTS should be viewed as opaque ob-

jects, unless the student is sufficiently expert to be able
to remove the pigment matter and so make them transparent. This can be done with moderate case in the
large compound eyes of the house-fly, but with most
other insects it is a delicate operation. The eyes are removed with sharp scissors, soaked in solution of caustic
potash, and the pigment washed from the inside, if necessary, with a fine wet camel's-hair brush. The work is
best done under water, and the cleaned eye may then be
examined in water. The compound eyes of insects like
the house-fly are usually so large and convex that only a
small part of the surface can be brought into focus at
the same time; they are often sufficiently transparent
to be quite easily studied without the troublesome
cleaning process. The facets of compound eyes often
have long hairs between them. The eyes of spiders are
several, and are arranged on the top of the head. If the
spider is small, place the whole body under a low power,
and view as an opaque object; otherwise cut into parts
to suit. The eyes have a wicked look as they brightly
gleam in the reflected light.

5. PROBOSCES OF INSECTS are as numerous and varied in
structure as the insects. They should be cut off close to
the body, squeezing the insect to force out the parts if
necessary. Examine in water or glycerine, covering them
with a thin glass square or circle as already explained.
The tongue or proboscis of a butterfly is interesting.
The latter is usually quite visible as a small coil just
under the front of the head. It may be unrolled with

a needle, and then cut off close to its point of attachment. It is composed of two longitudinal parts which separate easily. Examine in water or Canada balsam, and notice the one or more rows of little projections forming conspicuous barrel-shaped objects on the tip of some butterflies' tongues. These are supposed to be either papillæ of taste or implements with which to tear open the nectar glands within the flower.

6. Eggs of Insects cannot always be obtained when wanted, but they make beautiful objects. They should be viewed as opaque bodies, unless they are found after the larva has escaped, when they may be transparent and can be examined in water. The surface markings are in a great variety of patterns.

7. Insects, if small, are to be examined alive and as opaque objects. They may be imprisoned in a deep cell with a thin cover above them. If the cover tends to slide off when the microscope is inclined, allow a very small drop of water to run under one corner; the capillary attraction will then hold it in place. The quantity of water must not be sufficient to extend into the cell. If the insect is very small it may be killed by an immersion in a very strong solution of carbolic acid, and then transferred to Canada balsam, as described in the books devoted to microscopical mounting. This treatment often renders the body beautifully transparent, forcing out the probosces and spreading the legs in a very satisfactory manner.

8. Gizzards.—Pull off the head of a common cricket

or katydid, and the gizzard will usually be obtained as
an oblong, hard, reddish-brown enlargement attached
between the œsophagus and the upper part of the alimentary tube below the larger food-sack.   Free it from
these, cut it open lengthwise, and wash the interior with
a camel's-hair brush.   View it as a transparent object.
The grinding-teeth are well worth studying.   The gizzards of some beetles are also interesting.   They, however, are to be dissected out with fine scissors and needles.   The work is best done under water.

9. SCALES OF FISH are obtainable in great variety, and
make beautiful objects for low powers.   They are often
coated with a tenacious mucous material which may be
removed by washing in a solution of caustic potassa.
They should be examined in water or glycerine.

10. HAIR.—The hair of animals, from the human
animal down to the insects, forms an important class of
objects for study.   Every animal has characteristic hair
that may be recognized if the observer have sufficient
skill in the use of the microscope and sufficient knowledge of the subject; but to attempt to identify hair demands careful work with high powers.   It cannot be
done with a pocket-lens, nor with a one-inch objective,
unless the specimen is extremely common, and the observer has examined it often and attentively.   The difficulty is increased by the fact that the hair of our small
mammals differs greatly in microscopic structure in different parts of the body.   I once heard a lecturer claim
to have identified an unknown sample of hair with a

Coddington lens, a claim that he would have hesitated to make if he had known anything about the subject. A single fibre of hair can be recognized as such, and can be very readily distinguished from a fibre of wool, silk, or cotton, all of which the beginner should examine, since they will very often be found on the slide and may be mistaken. Hair is easily accessible on the cat, mouse, horse, and other animals. It should be examined in water with a high power. The hair of the mouse is very peculiar, as is also that of most bats. Colored fibres of wool are often seen on a slide, having dropped there from the observer's clothing or floated from the carpet. They must not be mistaken for a remarkable class of worms.

11. HAIRS OF ANTHRENUS LARVA.—The "buffalo beetle" (*Anthrénus scrophulárix*, and its relative, *A. museorum*) are pests that are becoming much too common in our houses. The larvæ, however, can be utilized by the microscopist as a supply for hairs of great interest. The reddish - brown tufts on the various parts of the body, especially of *Anthrenus museorum*, are composed of hairs having a remarkable structure, which the beginner is advised to examine for himself, especially since words cannot adequately describe it. The hairs should be studied in water or Canada balsam.

12. LINGUAL RIBBONS OF MOLLUSKS.—The tongue, palate, lingual ribbon, or odontophore (it has received several names) is the organ by which the water-snails rasp off their food from submerged stones and plants. It is

13*

a band-like body having many rows of short, variously formed teeth, which, under the microscope, present a beautiful and dazzling appearance. It is situated so that it may be partially protruded and used as a file, which not only removes the food particles, but carries them into the mouth. It may be obtained from any of the pond-snails, either by dissecting it out under the microscope with needles—a tedious process, successful in the hands of advanced workers—or by dissolving the soft parts of the snail by boiling the animal in a solution of caustic potassa. Drop the snail into a test-tube containing a small quantity of the solution, and boil until all of the soft body has disappeared. When cold pour off the liquid very gently, and the lingual ribbon will be found at the bottom usually without trouble, although it is often small and always perfectly transparent. The last few drops may be poured on a slip and examined with a low power, when the lingual ribbon will be seen and can be isolated with a needle. The form, size, and arrangement of the teeth vary greatly in the different water-snails. To see them is worth all the trouble that their preparation seems to demand. The ribbons should be examined in water or glycerine. A high power will often be required to show the teeth well.

13. The delicate epidermis from the leaves of plants can be obtained at any time in an unlimited supply and variety—in summer from the wild plants, in winter from those cultivated in the house. The cells forming this thin and usually colorless membrane vary in form and

size in the leaves of every plant, and the stomata or breathing pores also present very varied shapes, being most abundant on the under surface. The cells are usually empty; occasionally, however, they contain some chlorophyl grains and sometimes diffused coloring matter. The cuticle can be sliced off with a sharp razor, but a better way is to strip it off by tearing the leaf in a manner easily acquired but hardly describable. The piece obtained may be very small, but it will probably be sufficient for examination. All cuticles should be examined in water.

14. DÉUTZIA SCÁBRA, a handsome shrub now very common in cultivation, has leaves of special interest on account of the beautiful stellate hairs studding the surface. The cuticle may be removed and viewed as a transparent object, or a portion of the leaf may be examined by reflected light. By either method the observer will be pleased and interested. These stellate hairs are hard, brittle, and glass-like, and they are occasionally so well developed that they are visible to the naked eye as minute glistening stars. Under a good pocket-lens they are at all times apparent. They are most abundant on the upper surface.

15. OLEANDER LEAVES (*Nérium Oleánder*, a shrub frequently cultivated) have large stomata, which contain beneath the general surface of the leaf, but often projecting from the aperture of the breathing-pore, many short, variously curved hairs. To observe them, cut with a sharp razor a very thin slice across the leaf,

and examine the section in water, decolorizing and staining it, if desired, by some of the many processes described in the books devoted to microscopical mounting. The leaves also contain numerous crystals in the form called sphæraphides, a sphere of crystal roughened by minute points projecting from all parts of the surface.

16. PLANT HAIRS are as inexhaustible in appearance and structure as are animal hairs. They are formed of cells of varied size, shape, and contents. In many a circulation of the protoplasm or cyclosis may be noticed. They are simple or branched, terminating in a long point, a blunt apex, or in a spherical or other shaped gland. They may be examined by stripping off the cuticle and studying it as a transparent object in water. If the air adheres to the hairs and obscures them, as often happens to the very abundant branching hairs of the common mullein (*Verbáscum*), dip the cuticle in a drop of alcohol, and immediately transfer it to the drop of water already prepared on the slide. Hairs are to be found on some part of almost every plant.

17. POLLEN from the anthers of blooming flowers is one of the most attractive of very common objects which the beginner can find for examination as a dry mount. It can be obtained by simply tapping the slide with the ripe anther, when the pollen will be visible to the naked eye as a yellow dust, resolvable by the microscope into golden grains of varied forms and often of remarkable surface sculpturing. It, of course, differs in

shape, size, and appearance in different plants. The supply of new forms is therefore almost unlimited. In some the yellow dust consists of ovate grains with a central longitudinal depression, like a grain of wheat; in others it is triangular or spherical; in most it is delicately roughened or attractively marked. The pollen of the Passion-flower, the hollyhock, and the dandelion are especially noteworthy. The study of pollen and the drawing of the magnified image should be particularly pleasant work for ladies. The botanical study necessary to identify the plant supplying the pollen is advantageous and agreeable. The delicate microscopical work needed is more than pleasant, and is suited to the refined tastes of the ladies. The use of a pencil to record and preserve the beautiful forms and their markings will add much to the enjoyment and the profit; the work will then be both attractive and inspiring.

18. Seeds of wild plants form another almost inexhaustible group, some of them being exquisite beyond description. Even so common and so lowly a plant as the carrot has seeds of very peculiar appearance. The poppy, the cardinal flower (*Lobélia cardinális*), and other lobelias, are among the many worth noting. The portulaca and the many wild geraniums are also desirable. They should be examined dry as opaque objects. The seeds of *Collómia* have long been favorites on account of the peculiar spiral vessels on their surface. The plant is a subtropical one, and, so far as I know, does not grow uncultivated in any part of our country.

The seeds, however, are usually on sale by the microscopical dealers. To see the spirals, cut off a small piece of the outer coating of the seed, place it in a shallow cell with a thin cover above it. Arrange the slide on the stage, focus the objective, and while looking through the instrument allow a drop of water to run into the cell and around the object. Immediately the spiral vessels will seem to be springing and growing out of the seed in a remarkable way. They are adherent by means of a mucilaginous substance, soluble in water, and at the touch of the drop which dissolves their bonds they are set free, to the astonishment of the observer who sees them for the first time.

19. EQUISETUM SPORES (*Equisétum arvénse*) are worth collecting and examining as transparent dry objects. The plant is often called "Horse-tail" or "Scouring-rush," and is to be found almost everywhere in sterile places, especially along the railroad. The spores are small, almost spherical, and have four long narrow filaments which, when moistened, curl and curve and twist, and toss the spores about in every direction by what has been styled a "quadrupedal hornpipe." If a quantity of them is placed on a slide and gently breathed upon, the moisture of the breath will set those four long threads into motion, and the dance will at once begin. One of my friends makes a permanent mount of these curious spores by forming a cell of a single strand of gold-lace fringe, and of course covering them with a thin glass, fastened on with shellac cement, but so as not

to close the little apertures between the fibres of the
gold lace.   By breathing through these little openings
the moisture will at any time start the spores on their
"quadrupedal hornpipe," and when they are dry the
performance may be repeated.   An ordinary cell with
the cement ring cut into narrow parts by the strokes of
a penknife, so that the moisture could enter, would probably answer the purpose as well as my friend's more
elaborate contrivance.

20. PLANT CRYSTALS, already referred to (No. 15), are
found in the tissues of many plants. They occur in
four different and easily recognizable forms as follows :

a. Ráphides.—Small needle-like crystals, with long
shafts very gradually tapering to the pointed ends.
They are usually found collected together in loose bun-
dles, and generally within a distinct cell.   They are
abundant in *Lémna*, in common Spiderwort (*Trades-
cántia*), Touch-me-not (*Impátiens fúlva*), the Primrose
(*Œnothéra*), the Golden-club (*Oróntium aquáticum*),
Virginia creeper (*Ampelópsis quinquefólia*), and many
others.

b. Sphœráphides.—More or less spherical forms,
smooth, or with the entire surface roughened by crystalline projections.   They usually occur within a distinct
cell, and are to be found in the Oleander (*Nérium*), *Ge-
ranium, Óxalis*, Bouncing Bet (*Saponária officinális*),
Fleabane (*Erígeron*), *Portulaca, Hibiscus*, common mal-
low (*Málva rotúndifolia*), and others. Very remarkable crystals are found in the epidermal cells of the stem

of the common Richweed (*Pilea púmila*), a semitransparent plant growing in moist, shady places.

c. Long Crystal Prisms.—These are long crystals,
with angular, prismatic shafts and angular tips. They
never occur loosely as do Raphides, but one or two
together in the tissue of the plant. They are abundant
in the cultivated *Gladiolus*, Flower-de-luce (*Iris versi-
color*), some of the Asters, *Cýnthia Virgínica*, Hawk-
weed (*Hierácium*), the Thistles (*Cirsium*), and others.

d. Short Crystal Prisms.—These are cubical crystals,
long or short squares or prisms, indeed all those forms
which cannot be classed in the other divisions. They
are usually found in distinct cells, and often also in extensive rows or chains along the veins of leaves, especially in the Leguminosæ. They are abundant in the
Maple, Linden, white and red clover (*Trifólium répens,
T. praténse*), Onion, Monkey-flower (*Mímulus ríngens*),
Rabbit's-foot (*Trifólium arvénse*), and many other common plants.

When searching for these crystals a small fragment of
the plant should be crushed with a penknife, and examined in water with a moderately high power, as most
of the crystals are small. The cuticle should also be
stripped off. This may be done in the onion bulb and
Richweed (*Pilea*).

21. Crystals.—If the student has a polariscope he
will especially appreciate the beauty of crystals as ex-
emplified in color; if he has none he can study and ad-
mire the beauty of their forms. Almost any soluble salt

may be made to crystallize by preparing a strong solution and allowing it to slowly evaporate, and the formation of the crystals may be watched with the microscope. A small drop is placed on the slip and allowed to evaporate while on the stage. Sugar crystals can be prepared in this way. Common salt is very easily made to crystallize, and scarcely anything can be more beautiful than salt crystals viewed as opaque objects with a strong light reflected on and from them. The following are also noteworthy :*

Tartaric Acid.—Make a strong solution and place a large drop on the slide. Evaporate with a gentle heat by holding the slide several inches above the top of the lamp chimney.

Gallic Acid.—A small drop of a strong solution in alcohol should be allowed to evaporate very slowly.

Pyrogallic Acid.—A strong cold solution in water forms long needle-shaped crystals, " but if a very minute shower of some insoluble foreign substance be allowed to fall upon the solution when on the slide the effect is grand — each minute speck forming a nucleus around which the needle-shaped crystals gather, forming, if examined with a selenite slide, so resplendent an object that no words of mine can adequately describe it." *

Chlorate of Potash.—Make a strong solution in hot water and allow a small drop to spread evenly over the cell and evaporate slowly. To form dendritic or tree-

---

* *American Monthly Microscopical Journal*, August, 1883.

like crystals of this salt, heat a drop over the lamp. As soon as the crystals begin to form at any point, tilt the slide so that the liquor may run off, then continue the crystallization by gentle warmth.

There are many other salts which produce beautiful crystals when treated in the above or a similar manner, but the student would doubtless prefer to experiment for himself, rather than to have a bare list set down before him. And there are innumerable other common objects easily to be procured and worthy of study. It is not possible to enumerate a millionth part of them. Examine for yourself. Try and see what a good thing a microscope is. And the writer wishes the reader every success in the use of the delightful instrument.

GLOSSARY.

Acute : sharp or pointed.

Alimentary : pertaining to food.

Antenna (plural *antennæ*) *:* a jointed, movable tentacle or feeler on the head of certain crustacca and insects.

Anterior : front, going before.

Aquatic : living or growing in water.

Assimilated : turned to its own substance by digestion.

Beak : the lengthcned end or front.

Bi : in compound words, meaning two.

Bifid : two-parted.

Bosses : knobs, protuberances, usually rounded.

Branchial : relating to gills or branchiæ.

Carapace : the firm shell of some Iufusoria, Rotifers, etc.

Carnivorous : flesh-eating.

Caudal : pertaining to the tail.

Cellular : formed of or possessing cells.

Chlorophyl : the green coloring matter of plants.

Cilium (plural *cilia*) *:* a short, fine, vibrating hair.

Cœcum (plural *cœca*) *:* a part of the intestinal tube.

Colony : a cluster of several or many.

Comminution : the act of pulverizing or grinding.

Component : composing ; an elementary part.

Concave : hollow like a bowl.

Conical : cone-shaped.

Conjunction : union, association.

Constricted : suddenly narrowed or contracted.

Contractile : capable of being shortened or drawn together.

Cordate : heart-shaped.

Cornea : the transparent membrane forming the front of the eye.
Corpuscles : particles of matter.
Crenate : scalloped, or with rounded teeth.
Crescent : shaped like the new moon.
Crystalline : resembling crystal ; clear ; transparent.
Cuticle : the thin membrane covering the surface of plants ; the ●
   outermost layer of the skin.
Cyclosis : the movement of protoplasm within a closed cell.
Cylindrical : like a cylinder or long, circular body.

Dentate : toothed. .
Denticulate : with small, pointed teeth.
Diagonal : extending obliquely.
Diffused : spread out, extended.
Disintegrated : reduced to minute parts.
Distal : the furthest part.
Diverging : spreading from a central point.
Dorsal : pertaining to the back.
Dorsum : the back.

Ejected : thrown out.
Elliptical : oval.
Emarginate : notched.
Epithelium : the membrane lining various internal cavities and free
   surfaces of animals.
Expansile : capable of being expanded or widened.
Extensile : capable of being lengthened or extended.

Facet : a little surface or face.
Fascicle : a cluster.
Filament : a thread, or resembling a thread.
Fission : division or cleaving.
Flagellum (plural *flagella*) *:* a little lash.
Flexible : capable of being bent.
Frond : a leaf of fern; the entire plant of *Lemna*.
Frontal : pertaining to the front.
Frustule : the entire diatom, consisting of two *valves* and the *hoop*.
Furcate : forked.

Gelatinous : like jelly or gelatine.
Globule : a small round particle.
Granular : formed of or resembling small grains.
Granules : small grains.

Hemispherical : half a sphere.
Hexagon : a figure with six sides and angles.
Hispid : rough, with short, stiff hairs.
Homogeneous : of the same kind throughout.
Hyaline : glass-like, transparent.

Illoricate : without a lorica.
Imbricated : overlapping like shingles on a roof.
Invested : clothed, covered.

Labrum : a part of the mouth of crustacea and insects.
Lanceolate : lance-shaped.
Larva (plural *larvæ*)*:* an insect in its first stage after leaving the egg.
Laterally : by the sides.
Lophophore : the disk supporting the tentacles in the Polyzoa.
Lorica (plural *loricæ*)*:* the sheath or dwelling of certain microscopic
    animals.
Loricate : with a lorica.

Mastax : the internal jaws of the Rotifers.
Median : middle.
Membranous : formed of a thin skin.
Moniliform : like a string of beads.
Monograph : a treatise on a single subject.

Nodule : a small, rounded elevation.

Oblong : longer than broad.
Obtuse : blunt.
Œsophagus : the tubular passage extending from the pharynx or
    throat to the stomach.
Opaque : not transparent.
Ovoid : egg-shaped or oval.

Papilla (plural *papillæ*): a small rounded protuberance.

Parasite: a hanger on; as one animal or plant living at the expense of another.

Parietal: pertaining to the wall or side.

Pellet: a little ball or mass.

Pellucid: translucent or transparent.

Pendent: hanging.

Perianth: the leaves of a flower that cannot be distinguished into a calyx and corolla.

Pigment: coloring matter.

Podal: pertaining to, or used as, feet.

Polyp: a radiate animal, without locomotive organs, with retractile tentacles around the mouth, and a hollow body in which are suspended the digestive and other organs.

Posterior: the rear end.

Prehensile: adapted for grasping or seizing.

Process: a part prolonged or projecting beyond other parts connected with it.

Protoplasm: the semi-fluid, jelly-like contents of cells.

Protrusible: capable of being thrust forward.

Pulsating: throbbing, beating.

Recurved: directed backward.

Refracting: bending from a direct course.

Reticulated: with the form of a net.

Retort: a chemical glass vessel.

Retractile: capable of being drawn back or into the body.

Rudimental: imperfectly developed or formed; immature.

Segment: one of the rings or component parts of a worm or other body.

Semi: in compound words, meaning half.

Serrate: toothed like a saw.

Shaft: the stem or straight part between the ends.

Silicious: resembling or formed of silica.

Spherical: round like a ball.

Spinous: bearing spines.

Spiral: winding like a screw.

Spore: the minute seed of flowerless plants.

Statoblast: the winter egg of the Polyzoa.
Striated: finely streaked.
Sub: in compound words, meaning under, or less than.
Submerged: under water.
Sulcation: a groove.

Tortuous: winding, twisting.
Translucent: semitransparent.
Truncate: as if cut off square.
Tubercle: a small, knob-like elevation.
Tubular: resembling or formed of a tube.

Utricle: a little sack or bladder.

Ventral: pertaining to the lower surface; opposed to dorsal.
Ventrum: the concave side of *Closterium* (as here used).
Viscera: the intestines, or abdominal contents.
Visor: the fore-piece of a cap.

Whorl: several leaves in a circle around the stem.

Zoöphyte: a word applied to certain plant-like animals.

INDEX.

14

Cothurnia, 139, 146.
Craig microscope, 5.
Cricket, gizzard of, 281.
Cristatella, 225, 229, 231.
Crystals, plant, 289.
— polariscope, 290.
Cyclops, 241, 246, 254.
Cyclosis in *Anacharis*, 59.
— in *Closterium*, 77.
— in desmids, 68.
— in *Vallisneria*, 60.
Cynthia Virginica, crystal prisms in, 290.
Cyphoderia, 116, 129.
— *ampulla*, 129.
Cypris, 246, 250.

D.

Dandelion, 287.
Daphnia, 245, 247.
Davies, Thomas, on preparation and mounting of microscopic objects, 46.
Dendrocœlum lacteum, 183.
Dendromonas, 138, 140.
Dero, 164, 188, 192.
Desmidium, 74, 77.
— *Swartzii*, 77.
Desmids, 61, 64, 72.
— key to genera of, 66.
— to preserve, 88.
— vacuoles of, 68.
Deutzia scabra, 285.
Diaphragm of microscope stand, 22.
Diaptomus, 245, 246, 252.
Diatoma, 93, 94.
— *vulgare*, 94.
Diatoms, 61, 64, 72, 89, 92.
— as food for microscopic animals, 92.
Diatoms, fossil, 91.
— key to genera of, 93.
— literature, 92.
— movements, 66.
— structure, 90.
— surface markings, 67, 89.
— to study, 92.
Didymoprium, 74, 76.
— *Grevillii*, 76.
Difflugia, 112, 116, 123.
— *acuminata*, 125.

Difflugia corona, 125.
— *globulosa*, 125.
— *pyriformis*, 125.
Dinobryon, 138, 144.
Dinocharis, 208, 218.
Diplodontus, 267, 269, 273.
Dipper, tin, a useful collecting tool, 48.
Dipping-tube, 34.
— to make, 36.
— to use, 35.
Docidium, 75, 84.
— *Baculum*, 84.
— *crenulatum*, 84.
Draparnaldia, 104, 108.
— *glomerata*, 108.
Drawing the object, 41.
— camera lucida for, 42.
— reflector for, 42.
Draw-tube, microscope, 11.
Dry objects, to examine, 26.
— — to mount, 33.
Duckmeat (*Lemna*), 48, 57.

E.

Eggs of *Chironomus*, 167.
— of *Entomostraca*, 240.
— of insects, 281.
— of Rotifers, 205.
— of snails, 167.
— of snapping-turtle, 205.
— of *Tubifex*, 198.
— of *Turbellaria*, 181.
— of water-mites, 167.
Elytra, 279.
Emerton's "Life on the Sea-shore," x.
— "Structure and Habits of Spiders," x.
Enchytræus, 188, 189.
— *socialis*, 190.
— *vermicularis*, 190.
Encyonema, 93, 96.
— *paradoxa*, 96.
Entomostraca and Phyllopoda, 238.
— — key to genera of, 245.
— antennæ, 239.
— beak, 239.
— effect of heat on, 243.
— eggs of, 240.
— eyes of, 240.

THE END.

VALUABLE AND INTERESTING WORKS

FOR

PUBLIC & PRIVATE LIBRARIES,

PUBLISHED BY HARPER & BROTHERS, NEW YORK.

☞ *For a full List of Books suitable for Libraries published by HARPER & BROTH-ERS, see HARPER'S CATALOGUE, which may be had gratuitously on application to the publishers personally, or by letter enclosing Ten Cents in postage stamps.*

☞ *HARPER & BROTHERS will send their publications by mail, postage prepaid, on receipt of the price.*

MACAULAY'S ENGLAND. The History of England from the Ac-cession of James II. By THOMAS BABINGTON MACAULAY. New Edition, from New Electrotype Plates. 5 vols., in a Box, 8vo, Cloth, with Paper Labels, Uncut Edges and Gilt Tops, $10 00; Sheep, $12 50; Half Calf, $21 25. Sold only in Sets. Cheap Edition, 5 vols., 12mo, Cloth, $2 50.

MACAULAY'S MISCELLANEOUS WORKS. The Miscellaneous Works of Lord Macaulay. From New Electrotype Plates. 5 vols., in a Box, 8vo, Cloth, with Paper Labels, Uncut Edges and Gilt Tops, $10 00; Sheep, $12 50; Half Calf, $21 25. Sold only in Sets.

HUME'S ENGLAND. History of England, from the Invasion of Julius Cæsar to the Abdication of James II., 1688. By DAVID HUME. New and Elegant Library Edition, from New Electrotype Plates. 6 vols., in a Box, 8vo, Cloth, with Paper Labels, Uncut Edges and Gilt Tops, $12 00; Sheep, $15 00; Half Calf, $25 50. Sold only in Sets. Popular Edition, 6 vols., in a Box, 12mo, Cloth, $3 00.

GIBBON'S ROME. The History of the Decline and Fall of the Ro-man Empire. By EDWARD GIBBON. With Notes by Dean MIL-MAN, M. GUIZOT, and Dr. WILLIAM SMITH. New Edition, from New Electrotype Plates. 6 vols., 8vo, Cloth, with Paper Labels, Uncut Edges and Gilt Tops, $12 00; Sheep, $15 00; Half Calf, $25 50. Sold only in Sets. Popular Edition, 6 vols., in a Box, 12mo, Cloth, $3 00; Sheep, $6 00.

GOLDSMITH'S WORKS. The Works of Oliver Goldsmith. Edited by PETER CUNNINGHAM, F.S.A. From New Electrotype Plates. 4 vols., 8vo, Cloth, Paper Labels, Uncut Edges and Gilt Tops, $8 00; Sheep, $10 00; Half Calf, $17 00.

MOTLEY'S DUTCH REPUBLIC. The Rise of the Dutch Republic. A History. By JOHN LOTHROP MOTLEY, LL.D., D.C.L. With a Portrait of William of Orange. Cheap Edition, 3 vols., in a Box. 8vo, Cloth, with Paper Labels, Uncut Edges and Gilt Tops, $6 00; Sheep, $7 50; Half Calf, $12 75. Sold only in Sets. Original Library Edition, 3 vols., 8vo, Cloth, $10 50.

MOTLEY'S UNITED NETHERLANDS. History of the United Netherlands: From the Death of William the Silent to the Twelve Years' Truce—1584-1609. With a full View of the English-Dutch Struggle against Spain, and of the Origin and Destruction of the Spanish Armada. By JOHN LOTHROP MOTLEY, LL.D., D.C.L. Portraits. Cheap Edition, 4 vols., in a Box, 8vo, Cloth, with Paper Labels, Uncut Edges and Gilt Tops, $8 00; Sheep, $10 00; Half Calf, $17 00. Sold only in Sets. Original Library Edition, 4 vols., 8vo, Cloth, $14 00.

MOTLEY'S JOHN OF BARNEVELD. The Life and Death of John of Barneveld, Advocate of Holland. With a View of the Primary Causes and Movements of the "Thirty Years' War." By JOHN LOTHROP MOTLEY, LL.D., D.C.L. Illustrated. Cheap Edition, 2 vols., in a Box, 8vo, Cloth, with Paper Labels, Uncut Edges and Gilt Tops, $4 00; Sheep, $5 00; Half Calf, $8 50. Sold only in Sets. Original Library Edition, 2 vols., 8vo, Cloth, $7 00.

HILDRETH'S UNITED STATES. History of the United States. FIRST SERIES: From the Discovery of the Continent to the Organization of the Government under the Federal Constitution. SECOND SERIES: From the Adoption of the Federal Constitution to the End of the Sixteenth Congress. By RICHARD HILDRETH. Popular Edition, 6 vols., in a Box, 8vo, Cloth, with Paper Labels, Uncut Edges and Gilt Tops, $12 00; Sheep, $15 00; Half Calf, $25 50. Sold only in Sets.

LODGE'S ENGLISH COLONIES IN AMERICA. English Colonies in America. A Short History of the English Colonies in America. By HENRY CABOT LODGE. New and Revised Edition. 8vo, Half Leather, $3 00.

TREVELYAN'S LIFE OF MACAULAY. The Life and Letters of Lord Macaulay. By his Nephew, G. OTTO TREVELYAN, M.P. With Portrait on Steel. 2 vols., 8vo, Cloth, Uncut Edges and Gilt Tops, $5 00; Sheep, $6 00; Half Calf, $9 50. Popular Edition, 2 vols. in one, 12mo, Cloth, $1 75.

TREVELYAN'S LIFE OF FOX. The Early History of Charles James Fox. By GEORGE OTTO TREVELYAN. 8vo, Cloth, Uncut Edges and Gilt Tops, $2 50; Half Calf, $4 75.

WRITINGS AND SPEECHES OF SAMUEL J. TILDEN. Edited by JOHN BIGELOW. 2 vols., 8vo, Cloth, Gilt Tops and Uncut Edges, $6 00 per set.

GENERAL DIX'S MEMOIRS. Memoirs of John Adams Dix. Compiled by his Son, MORGAN DIX. With Five Steel-plate Portraits. 2 vols., 8vo, Cloth, Gilt Tops and Uncut Edges, $5 00.

HUNT'S MEMOIR OF MRS. LIVINGSTON. A Memoir of Mrs. Edward Livingston. With Letters hitherto Unpublished. By LOUISE LIVINGSTON HUNT. 12mo, Cloth, $1 25.

GEORGE ELIOT'S LIFE. George Eliot's Life, Related in her Letters and Journals. Arranged and Edited by her Husband, J. W. CROSS. Portraits and Illustrations. In Three Volumes. 12mo, Cloth, $3 75. New Edition, with Fresh Matter. (Uniform with "Harper's Library Edition" of George Eliot's Works.)

PEARS'S FALL OF CONSTANTINOPLE. The Fall of Constantinople. Being the Story of the Fourth Crusade. By EDWIN PEARS, LL.B. 8vo, Cloth, $2 50.

RANKE'S UNIVERSAL HISTORY. The Oldest Historical Group of Nations and the Greeks. By LEOPOLD VON RANKE. Edited by G. W. PROTHERO, Fellow and Tutor of King's College, Cambridge. Vol. I. 8vo, Cloth, $2 50.

LIFE AND TIMES OF THE REV. SYDNEY SMITH. A Sketch of the Life and Times of the Rev. Sydney Smith. Based on Family Documents and the Recollections of Personal Friends. By STUART J. REID. With Steel-plate Portrait and Illustrations. 8vo, Cloth, $3 00.

STORMONTH'S ENGLISH DICTIONARY. A Dictionary of the English Language, Pronouncing, Etymological, and Explanatory: embracing Scientific and other Terms, Numerous Familiar Terms, and a Copious Selection of Old English Words. By the Rev. JAMES STORMONTH. The Pronunciation Revised by the Rev. P. H. PHELP, M.A. Imperial 8vo, Cloth, $6 00; Half Roan, $7 00; Full Sheep, $7 50. (New Edition.)

PARTON'S CARICATURE. Caricature and Other Comic Art, in All Times and Many Lands. By JAMES PARTON. 203 Illustrations. 8vo, Cloth, Uncut Edges and Gilt Tops, $5 00; Half Calf, $7 25.

DU CHAILLU'S LAND OF THE MIDNIGHT SUN. Summer and Winter Journeys in Sweden, Norway, Lapland, and Northern Finland. By PAUL B. DU CHAILLU. Illustrated. 2 vols., 8vo, Cloth, $7 50; Half Calf, $12 00.

LOSSING'S CYCLOPÆDIA OF UNITED STATES HISTORY.
From the Aboriginal Period to 1876. By B. J. LOSSING, LL.D.
Illustrated by 2 Steel Portraits and over 1000 Engravings. 2 vols.,
Royal 8vo, Cloth, $10 00; Sheep, $12 00; Half Morocco, $15 00.
(*Sold by Subscription only.*)

LOSSING'S FIELD-BOOK OF THE REVOLUTION. Pictorial
Field-Book of the Revolution; or, Illustrations by Pen and Pencil
of the History, Biography, Scenery, Relics, and Traditions of the
War for Independence. By BENSON J. LOSSING. 2 vols., 8vo,
Cloth, $14 00; Sheep or Roan, $15 00; Half Calf, $18 00.

LOSSING'S FIELD-BOOK OF THE WAR OF 1812. Pictorial
Field-Book of the War of 1812; or, Illustrations by Pen and Pencil
of the History, Biography, Scenery, Relics, and Traditions of the
last War for American Independence. By BENSON J. LOSSING.
With several hundred Engravings. 1088 pages, 8vo, Cloth, $7 00;
Sheep or Roan, $8 50; Half Calf, $10 00.

**MÜLLER'S POLITICAL HISTORY OF RECENT TIMES (1816–
1875).** With Special Reference to Germany. By WILLIAM MÜL-
LER. Translated, with an Appendix covering the Period from 1876
to 1881, by the Rev. JOHN P. PETERS, Ph.D. 12mo, Cloth, $3 00.

STANLEY'S THROUGH THE DARK CONTINENT. Through
the Dark Continent; or, The Sources of the Nile, Around the Great
Lakes of Equatorial Africa, and Down the Livingstone River to the
Atlantic Ocean. 149 Illustrations and 10 Maps. By H. M. STAN-
LEY. 2 vols., 8vo, Cloth, $10 00; Sheep, $12 00; Half Morocco,
$15 00.

STANLEY'S CONGO. The Congo and the Founding of its Free
State, a Story of Work and Exploration. With over One Hundred
Full-page and smaller Illustrations, Two Large Maps, and several
smaller ones. By H. M. STANLEY. 2 vols., 8vo, Cloth, $10 00;
Sheep, $12 00; Half Morocco, $15 00.

GREEN'S ENGLISH PEOPLE. History of the English People.
By JOHN RICHARD GREEN, M.A. With Maps. 4 vols., 8vo, Cloth,
$10 00; Sheep, $12 00; Half Calf, $19 00.

GREEN'S MAKING OF ENGLAND. The Making of England.
By JOHN RICHARD GREEN. With Maps. 8vo, Cloth, $2 50; Sheep,
$3 00; Half Calf, $3 75.

GREEN'S CONQUEST OF ENGLAND. The Conquest of England.
By JOHN RICHARD GREEN. With Maps. 8vo, Cloth, $2 50; Sheep,
$3 00; Half Calf, $3 75.

ENGLISH MEN OF LETTERS. Edited by JOHN MORLEY. The following volumes are now ready. Others will follow:

JOHNSON. By I. Stephen.—GIBBON. By J. C. Morison.—SCOTT. By R. H. Hutton.—SHELLEY. By J. A. Symonds.—GOLDSMITH. By W. Black.—HUME. By Professor Huxley.—DEFOE. By W. Minto.—BURNS. By Principal Shairp.—SPENSER. By R. W. Church.—THACKERAY. By A. Trollope.—BURKE. By J. Morley.—MILTON. By M. Pattison.—SOUTHEY. By E. Dowden.—CHAUCER. By A. W. Ward.—BUNYAN. By J. A. Froude.—COWPER. By G. Smith.—POPE. By L. Stephen.—BYRON. By J. Nichols.—LOCKE. By T. Fowler.—WORDSWORTH. By F. W. H. Myers.—HAWTHORNE. By Henry James, Jr.—DRYDEN. By G. Saintsbury.—LANDOR. By S. Colvin.—DE QUINCEY. By D. Masson.—LAMB. By A. Ainger.—BENTLEY. By R. C. Jebb.—DICKENS. By A. W. Ward.—GRAY. By E. W. Gosse.—SWIFT. By L. Stephen. —STERNE. By H. D. Traill.—MACAULAY. By J. C. Morison.—FIELDING. By A. Dobson.—SHERIDAN. By Mrs. Oliphant.—ADDISON. By W. J. Courthope.—BACON. By R. W. Church.—COLERIDGE. By H. D. Traill.—SIR PHILIP SIDNEY. By J. A. Symonds. 12mo, Cloth, 75 cents per volume.

REBER'S HISTORY OF ANCIENT ART. History of Ancient Art. By Dr. FRANZ VON REBER. Revised by the Author. Translated and Augmented by Joseph Thacher Clarke. With 310 Illustrations and a Glossary of Technical Terms. 8vo, Cloth, $3 50.

REBER'S MEDIÆVAL ART. History of Mediæval Art. By Dr. FRANZ VON REBER. Translated and Augmented by Joseph Thacher Clarke. With 422 Illustrations, and a Glossary of Technical Terms. 8vo, Cloth, $5 00.

NEWCOMB'S ASTRONOMY. Popular Astronomy. By SIMON NEWCOMB, LL.D. With 112 Engravings, and 5 Maps of the Stars. 8vo, Cloth, $2 50; School Edition, 12mo, Cloth, $1 30.

VAN-LENNEP'S BIBLE LANDS. Bible Lands: their Modern Customs and Manners Illustrative of Scripture. By HENRY J. VAN-LENNEP, D.D. 350 Engravings and 2 Colored Maps. 8vo, Cloth, $5 00; Sheep, $6 00; Half Morocco, $8 00.

CESNOLA'S CYPRUS. Cyprus: its Ancient Cities, Tombs, and Temples. A Narrative of Researches and Excavations during Ten Years' Residence in that Island. By L. P. DI CESNOLA. With Portrait, Maps, and 400 Illustrations. 8vo, Cloth, Extra, Uncut Edges and Gilt Tops, $7 50.

TENNYSON'S COMPLETE POEMS. The Complete Poetical Works of Alfred, Lord Tennyson. With an Introductory Sketch by Anne Thackeray Ritchie. With Portraits and Illustrations. 8vo, Extra Cloth, Bevelled, Gilt Edges, $2 50.

SHORT'S NORTH AMERICANS OF ANTIQUITY. The North Americans of Antiquity. Their Origin, Migrations, and Type of Civilization Considered. By JOHN T. SHORT. Illustrated. 8vo, Cloth, $3 00.

GROTE'S HISTORY OF GREECE. 12 vols., 12mo, Cloth, $18 00; Sheep, $22 80; Half Calf, $39 00.

FLAMMARION'S ATMOSPHERE. Translated from the French of CAMILLE FLAMMARION. With 10 Chromo-Lithographs and 86 Wood-cuts. 8vo, Cloth, $6 00; Half Calf, $8 25.

BAKER'S ISMAÏLIA : a Narrative of the Expedition to Central Africa for the Suppression of the Slave-trade, organized by Ismaïl, Khedive of Egypt. By Sir SAMUEL W. BAKER. With Maps, Portraits, and Illustrations. 8vo, Cloth, $5 00; Half Calf, $7 25.

LIVINGSTONE'S ZAMBESI. Narrative of an Expedition to the Zambesi and its Tributaries, and of the Discovery of the Lakes Shirwa and Nyassa, 1858 to 1864. By DAVID and CHARLES LIVINGSTONE. Illustrated. 8vo, Cloth, $5 00; Sheep, $5 50; Half Calf, $7 25.

LIVINGSTONE'S LAST JOURNALS. The Last Journals of David Livingstone, in Central Africa, from 1865 to his Death. Continued by a Narrative of his Last Moments, obtained from his Faithful Servants Chuma and Susi. By HORACE WALLER. With Portrait, Maps, and Illustrations. 8vo, Cloth, $5 00; Sheep, $6 00.

BLAIKIE'S LIFE OF DAVID LIVINGSTONE. Memoir of his Personal Life, from his Unpublished Journals and Correspondence. By W. G. BLAIKIE, D.D. With Portrait and Map. 8vo, Cloth, $2 25.

"THE FRIENDLY EDITION" of Shakespeare's Works. Edited by W. J. ROLFE. In 20 vols. Illustrated. 16mo, Gilt Tops and Uncut Edges, Sheets, $27 00; Cloth, $30 00; Half Calf, $60 per Set.

GIESELER'S ECCLESIASTICAL HISTORY. A Text-Book of Church History. By Dr. JOHN C. L. GIESELER. Translated from the Fourth Revised German Edition. Revised and Edited by Rev. HENRY B. SMITH, D.D. Vols. I., II., III., and IV., 8vo, Cloth, $2 25 each; Vol. V., 8vo, Cloth, $3 00. Complete Sets, 5 vols., Sheep, $14 50; Half Calf, $23 25.

CURTIS'S LIFE OF BUCHANAN. Life of James Buchanan, Fifteenth President of the United States. By GEORGE TICKNOR CURTIS. With Two Steel Plate Portraits. 2 vols., 8vo, Cloth, Uncut Edges and Gilt Tops, $6 00.

COLERIDGE'S WORKS. The Complete Works of Samuel Taylor Coleridge. With an Introductory Essay upon his Philosophical and Theological Opinions. Edited by Professor W. G. T. SHEDD. With Steel Portrait, and an Index. 7 vols., 12mo, Cloth, $2 00 per volume; $12 00 per set; Half Calf, $24 25.

GRIFFIS'S JAPAN. The Mikado's Empire: Book I. History of Japan, from 660 B.C. to 1872 A.D. Book II. Personal Experiences, Observations, and Studies in Japan, from 1870 to 1874. With Two Supplementary Chapters: Japan in 1883, and Japan in 1886. By W. E. GRIFFIS. Copiously Illustrated. 8vo, Cloth, $4 00; Half Calf, $6 25.

SMILES'S HISTORY OF THE HUGUENOTS. The Huguenots: their Settlements, Churches, and Industries in England and Ireland. By SAMUEL SMILES. With an Appendix relating to the Huguenots in America. Crown, 8vo, Cloth, $2 00.

SMILES'S HUGUENOTS AFTER THE REVOCATION. The Huguenots in France after the Revocation of the Edict of Nantes; with a Visit to the Country of the Vaudois. By SAMUEL SMILES. Crown 8vo, Cloth, $2 00.

SMILES'S LIFE OF THE STEPHENSONS. The Life of George Stephenson, and of his Son, Robert Stephenson; comprising, also, a History of the Invention and Introduction of the Railway Locomotive. By SAMUEL SMILES. Illustrated. 8vo, Cloth, $3 00.

THE POETS AND POETRY OF SCOTLAND: From the Earliest to the Present Time. Comprising Characteristic Selections from the Works of the more Noteworthy Scottish Poets, with Biographical and Critical Notices. By JAMES GRANT WILSON. With Portraits on Steel. 2 vols., 8vo, Cloth, $10 00; Gilt Edges, $11 00.

SCHLIEMANN'S ILIOS. Ilios, the City and Country of the Trojans. A Narrative of the Most Recent Discoveries and Researches made on the Plain of Troy. By Dr. HENRY SCHLIEMANN. Maps, Plans, and Illustrations. Imperial 8vo, Illuminated Cloth, $12 00; Half Morocco, $15 00.

SCHLIEMANN'S TROJA. Troja. Results of the Latest Researches and Discoveries on the Site of Homer's Troy, and in the Heroic Tumuli and other Sites, made in the Year 1882, and a Narrative of a Journey in the Troad in 1881. By Dr. HENRY SCHLIEMANN. Preface by Professor A. H. Sayce. With Wood-cuts, Maps, and Plans. 8vo, Cloth, $7 50; Half Morocco, $10 00.

SCHWEINFURTH'S HEART OF AFRICA. Three Years' Travels and Adventures in the Unexplored Regions of the Centre of Africa—from 1868 to 1871. By GEORG SCHWEINFURTH. Translated by ELLEN E. FREWER. Illustrated. 2 vols., 8vo, Cloth, $8 00.

NORTON'S STUDIES OF CHURCH-BUILDING. Historical Studies of Church-Building in the Middle Ages. Venice, Siena, Florence. By CHARLES ELIOT NORTON. 8vo, Cloth, $3 00.

THE VOYAGE OF THE "CHALLENGER." The Atlantic: an
Account of the General Results of the Voyage during 1873, and the
Early Part of 1876. By Sir WYVILLE THOMSON, K.C.B., F.R.S.
Illustrated.   2 vols., 8vo, Cloth, $12 00.

THE STUDENT'S SERIES. Maps and Illustrations.   12mo, Cloth:
FRANCE.—GIBBON.—GREECE.—ROME (by LIDDELL).—OLD TES-
TAMENT HISTORY. — NEW TESTAMENT HISTORY. — STRICKLAND'S
QUEENS OF ENGLAND.—ANCIENT HISTORY OF THE EAST.—HAL-
LAM'S MIDDLE AGES. — HALLAM'S CONSTITUTIONAL HISTORY OF
ENGLAND.— LYELL'S ELEMENTS OF GEOLOGY. — MERIVALE'S GEN-
ERAL HISTORY OF ROME.—COX'S GENERAL HISTORY OF GREECE.
—CLASSICAL DICTIONARY.—SKEAT'S ETYMOLOGICAL DICTIONARY.—
RAWLINSON'S ANCIENT HISTORY. $1 25 per volume.
LEWIS'S HISTORY OF GERMANY.—ECCLESIASTICAL HISTORY, Two
Vols.—HUME'S ENGLAND.—MODERN EUROPE. $1 50 per volume.
WESTCOTT AND HORT'S GREEK TESTAMENT, $1 00.

THOMSON'S SOUTHERN PALESTINE AND JERUSALEM.
Southern Palestine and Jerusalem.   Biblical Illustrations drawn
from the Manners and Customs, the Scenes and Scenery, of the
Holy Land.   By W. M. THOMSON, D.D.   140 Illustrations and
Maps.   Square 8vo, Cloth, $6 00; Sheep, $7 00; Half Morocco,
$8 50; Full Morocco, Gilt Edges, $10 00.

THOMSON'S CENTRAL PALESTINE AND PHŒNICIA.   Cen-
tral Palestine and Phoenicia.   Biblical Illustrations drawn from the
Manners and Customs, the Scenes and Scenery, of the Holy Land.
By W. M. THOMSON, D.D.   130 Illustrations and Maps.   Square 8vo,
Cloth, $6 00; Sheep, $7 00; Half Morocco, $8 50; Full Morocco,
$10 00.

THOMSON'S LEBANON, DAMASCUS, AND BEYOND JORDAN.
Lebanon, Damascus, and beyond Jordan.   Biblical Illustrations drawn
from the Manners and Customs, the Scenes and Scenery, of the Holy
Land.   By W. M. THOMSON, D.D.   147 Illustrations and Maps.
Square 8vo, Cloth, $6 00; Sheep, $7 00; Half Morocco, $8 50;
Full Morocco, $10 00.
Popular Edition of the above three volumes, 8vo, Ornamental Cloth,
$9 00 per set.

CYCLOPÆDIA OF BRITISH AND AMERICAN POETRY.   Edited by EPES SARGENT.   Royal 8vo, Illuminated Cloth, Colored
Edges, $4 50; Half Leather, $5 00.

EATON'S CIVIL SERVICE.   Civil Service in Great Britain.   A
History of Abuses and Reforms, and their bearing upon American
Politics.   By DORMAN B. EATON.   8vo, Cloth, $2 50.

CAMERON'S ACROSS AFRICA. Across Africa. By VERNEY LOV-
ETT CAMERON. Map and Illustrations. 8vo, Cloth, $5 00.

CARLYLE'S FREDERICK THE GREAT. History of Friedrich
II., called Frederick the Great. By THOMAS CARLYLE. Portraits,
Maps, Plans, &c. 6 vols., 12mo, Cloth, $7 50; Sheep, $9 90; Half
Calf, $18 00.

CARLYLE'S FRENCH REVOLUTION. The French Revolution:
a History. By THOMAS CARLYLE. 2 vols., 12mo, Cloth, $2 50;
Sheep, $2 90; Half Calf, $4 25.

CARLYLE'S OLIVER CROMWELL. Oliver Cromwell's Letters
and Speeches, including the Supplement to the First Edition. With
Elucidations. By THOMAS CARLYLE. 2 vols., 12mo, Cloth, $2 50;
Sheep, $2 90; Half Calf, $4 25.

PAST AND PRESENT, CHARTISM, AND SARTOR RESARTUS.
By THOMAS CARLYLE. 12mo, Cloth, $1 25.

EARLY KINGS OF NORWAY, AND THE PORTRAITS OF JOHN
KNOX. By THOMAS CARLYLE. 12mo, Cloth, $1 25.

REMINISCENCES BY THOMAS CARLYLE. Edited by J. A.
FROUDE. 12mo, Cloth, with Copious Index, and with Thirteen Portraits, 50 cents.

FROUDE'S LIFE OF THOMAS CARLYLE. PART I. A History
of the First Forty Years of Carlyle's Life (1795-1835). By JAMES
ANTHONY FROUDE, M.A. With Portraits and Illustrations. 2 volumes in one, 12mo, Cloth, $1 00.

    PART II. A History of Carlyle's Life in London (1834-1881). By
JAMES ANTHONY FROUDE. Illustrated. 2 volumes in one. 12mo,
Cloth, $1 00.

M'CARTHY'S HISTORY OF ENGLAND. A History of Our Own
Times, from the Accession of Queen Victoria to the General Elec-
tion of 1880. By JUSTIN M'CARTHY. 2 vols., 12mo, Cloth, $2 50;
Half Calf, $6 00.

M'CARTHY'S SHORT HISTORY OF OUR OWN TIMES. A
Short History of Our Own Times, from the Accession of Queen Victoria to the General Election of 1880. By JUSTIN M'CARTHY, M.P.
12mo, Cloth, $1 50.

M'CARTHY'S HISTORY OF THE FOUR GEORGES. A History
of the Four Georges. By JUSTIN M'CARTHY, M.P. Vol. I. 12mo,
Cloth, $1 25. (To be completed in Four Volumes.)

ABBOTT'S HISTORY OF THE FRENCH REVOLUTION. The French Revolution of 1789, as viewed in the Light of Republican Institutions. By JOHN S. C. ABBOTT. Illustrated. 8vo, Cloth, $5 00; Sheep, $5 50; Half Calf, $7 25.

ABBOTT'S NAPOLEON. The History of Napoleon Bonaparte. By JOHN S. C. ABBOTT. Maps, Illustrations, and Portraits. 2 vols., 8vo, Cloth, $10 00; Sheep, $11 00; Half Calf, $14 50.

ABBOTT'S NAPOLEON AT ST. HELENA. Napoleon at St. Helena; or, Anecdotes and Conversations of the Emperor during the Years of his Captivity. Collected from the Memorials of Las Casas, O'Meara, Montholon, Antommarchi, and others. By JOHN S. C. ABBOTT. Illustrated. 8vo, Cloth, $5 00; Sheep, $5 50; Half Calf, $7 25.

ABBOTT'S FREDERICK THE GREAT. The History of Frederick the Second, called Frederick the Great. By JOHN S. C. ABBOTT. Illustrated. 8vo, Cloth, $5 00; Half Calf, $7 25.

TROLLOPE'S AUTOBIOGRAPHY. An Autobiography. By ANTHONY TROLLOPE. With a Portrait. 12mo, Cloth, $1 25.

TROLLOPE'S CICERO. Life of Cicero. By ANTHONY TROLLOPE. 2 vols., 12mo, Cloth, $3 00.

FOLK-LORE OF SHAKESPEARE. By the Rev. T. F. THISELTON DYER, M.A., Oxon. 8vo, Cloth, $2 50.

WATSON'S MARCUS AURELIUS ANTONINUS. Marcus Aurelius Antoninus. By PAUL BARRON WATSON. Crown 8vo, Cloth, $2 50.

THOMSON'S THE GREAT ARGUMENT. The Great Argument; or, Jesus Christ in the Old Testament. By W. H. THOMSON, M.A., M.D. Crown 8vo, Cloth, $2 00.

HUDSON'S HISTORY OF JOURNALISM. Journalism in the United States, from 1690 to 1872. By FREDERIC HUDSON. 8vo, Cloth, $5 00; Half Calf, $7 25.

SHELDON'S HISTORY OF CHRISTIAN DOCTRINE. History of Christian Doctrine. By H. C. SHELDON, Professor of Church History in Boston University. 2 vols., 8vo, Cloth, $3 50 per set.

DEXTER'S CONGREGATIONALISM. The Congregationalism of the Last Three Hundred Years, as Seen in its Literature: with Special Reference to certain Recondite, Neglected, or Disputed Passages. With a Bibliographical Appendix. By H. M. DEXTER. Large 8vo, Cloth, $6 00.

SYMONDS'S SKETCHES AND STUDIES IN SOUTHERN EUROPE. By JOHN ADDINGTON SYMONDS. 2 vols., Square 16mo, Cloth, $4 00; Half Calf, $7 50.

SYMONDS'S GREEK POETS. Studies of the Greek Poets. By JOHN ADDINGTON SYMONDS. 2 vols., Square 16mo, Cloth, $3 50; Half Calf, $7 00.

MAHAFFY'S GREEK LITERATURE. A History of Classical Greek Literature. By J. P. MAHAFFY. 2 vols., 12mo, Cloth, $4 00; Half Calf, $7 50.

DU CHAILLU'S ASHANGO LAND. A Journey to Ashango Land, and Further Penetration into Equatorial Africa. By PAUL B. DU CHAILLU. Illustrated. 8vo, Cloth, $5 00; Half Calf, $7 25.

SIMCOX'S LATIN LITERATURE. A History of Latin Literature, from Ennius to Boethius. By GEORGE AUGUSTUS SIMCOX, M.A. 2 vols., 12mo, Cloth, $4 00.

BARTLETT'S FROM EGYPT TO PALESTINE. Through Sinai, the Wilderness, and the South Country. Observations of a Journey made with Special Reference to the History of the Israelites. By S. C. BARTLETT, D.D. Maps and Illustrations. 8vo, Cloth, $3 50.

KINGLAKE'S CRIMEAN WAR. The Invasion of the Crimea: its Origin, and an Account of its Progress down to the Death of Lord Raglan. By ALEXANDER WILLIAM KINGLAKE. With Maps and Plans. Four Volumes now ready. 12mo, Cloth, $2 00 per vol.

NEWCOMB'S POLITICAL ECONOMY. Principles of Political Economy. By SIMON NEWCOMB, LL.D., Professor of Mathematics, U. S. Navy, Professor in the Johns Hopkins University. pp. xvi., 548. 8vo, Cloth, $2 50.

SHAKSPEARE. The Dramatic Works of Shakspeare. With Notes. Engravings. 6 vols., 12mo, Cloth, $9 00. 2 vols., 8vo, Cloth, $4 00; Sheep, $5 00. In one vol., 8vo, Sheep, $4 00.

GENERAL BEAUREGARD'S MILITARY OPERATIONS. The Military Operations of General Beauregard in the War Between the States, 1861 to 1865; including a brief Personal Sketch, and a Narrative of his Services in the War with Mexico, 1846 to 1848. By ALFRED ROMAN, formerly Aide-de-Camp on the Staff of General Beauregard. With Portraits, &c. 2 vols., 8vo, Cloth, $7 00; Sheep, $9 00; Half Morocco, $11 00; Full Morocco, $15 00. (*Sold only by Subscription.*)

NORDHOFF'S COMMUNISTIC SOCIETIES OF THE UNITED STATES. The Communistic Societies of the United States, from Personal Visit and Observation; including Detailed Accounts of the Economists, Zoarites, Shakers, the Amana, Oneida, Bethel, Aurora, Icarian, and other existing Societies. By CHARLES NORDHOFF. Illustrations. 8vo, Cloth, $4 00.

BOSWELL'S JOHNSON. The Life of Samuel Johnson, LL.D., including a Journal of a Tour to the Hebrides. By JAMES BOSWELL. Edited by J. W. CROKER, LL.D., F.R.S. With a Portrait of Boswell. 2 vols., 8vo, Cloth, $4 00; Sheep, $5 00.

BROUGHAM'S AUTOBIOGRAPHY. Life and Times of Henry, Lord Brougham. Written by Himself. 3 vols., 12mo, Cloth, $6 00.

BOURNE'S LOCKE. The Life of John Locke. By H. R. Fox BOURNE. 2 vols., 8vo, Cloth, $5 00.

BARTH'S NORTH AND CENTRAL AFRICA. Travels and Discoveries in North and Central Africa : being a Journal of an Expedition undertaken under the Auspices of H.B.M.'s Government, in the Years 1849–1855. By HENRY BARTH, Ph.D., D.C.L. Illustrated. 3 vols, 8vo, Cloth, $12 00.

BULWER'S LIFE AND LETTERS. Life, Letters, and Literary Remains of Edward Bulwer, Lord Lytton. By his Son, the EARL OF LYTTON ("Owen Meredith"). Volume I. Illustrated. 12mo, Cloth, $2 75.

BULWER'S HORACE. The Odes and Epodes of Horace. A Metrical Translation into English. With Introduction and Commentaries. With Latin Text from the Editions of Orelli, Macleane, and Yonge. 12mo, Cloth, $1 75.

BULWER'S MISCELLANEOUS WORKS. Miscellaneous Prose Works of Edward Bulwer, Lord Lytton. In Two Volumes. 12mo, Cloth, $3 50.

PERRY'S HISTORY OF THE CHURCH OF ENGLAND. A History of the English Church, from the Accession of Henry VIII. to the Silencing of Convocation. By G. G. PERRY, M.A. With a Sketch of the History of the Protestant Episcopal Church in the United States, by J. A. SPENCER, S.T.D. Crown 8vo, Cloth, $2 50.

FORSTER'S LIFE OF DEAN SWIFT. The Early Life of Jonathan Swift (1667–1711). By JOHN FORSTER. With Portrait. 8vo, Cloth, Uncut Edges and Gilt Tops, $2 50.

www.ingramcontent.com/pod-product-compliance
Lightning Source LLC
Chambersburg PA
CBHW021459210326
41599CB00012B/1058